Father to Son

Truth, Reason, and Decency

OTHER TITLES BY JAMES D. WATSON

Watson JD. 1968. *The Double Helix: A Personal Account of the Discovery of the Structure of DNA*. Atheneum, New York

Watson JD. 1968. Gunther S. Stent, ed. *The Double Helix: A Personal Account of the Discovery of the Structure of DNA*. W. W. Norton & Company, New York [Norton Critical Editions, 1981].

Watson JD. 2002. *Genes, Girls, and Gamow: After the Double Helix*. Random House, New York.

Watson JD. 2007. *Avoid Boring People and Other Lessons from a Life in Science*. Random House, New York.

Watson JD, Gann A, Witkowski JA. 2012. *The Annotated and Illustrated Double Helix*. Simon & Schuster, New York.

OTHER RELATED TITLES FROM COLD SPRING HARBOR LABORATORY PRESS

Friedberg EC. 2004. *The Writing Life of James D. Watson*.

Watson JD. 2001. *A Passion for DNA: Genes, Genomes, and Society*.

Cairns J, Stent GS, Watson JD. 2007. *Phage and the Origins of Molecular Biology: The Centennial Edition*.

Inglis JR, Sambrook J, Witkowski JA, eds. 2003. *Inspiring Science: Jim Watson and the Age of DNA*.

Father to Son

Truth, Reason, and Decency

JAMES D. WATSON

COLD SPRING HARBOR LABORATORY PRESS
Cold Spring Harbor, New York • www.cshlpress.org

Father to Son: Truth, Reason, and Decency

Published by Cold Spring Harbor Laboratory Press
Printed in the United States of America

Publisher	John Inglis
Developmental Editor	Judy Cuddihy
Director of Editorial Development	Jan Argentine
Project Manager	Inez Sialiano
Permissions Coordinators	Carol Brown, Maryliz Dickerson, Inez Sialiano
Production Editor	Kathleen Bubbeo
Production Manager	Denise Weiss
Cover Designer	Michael Albano

Front cover artwork: (*Clockwise from top left*) "Eisenhower" portrait of James Dewey Watson, Sr. (JDW, Sr.) in his 60s; JDW, Sr. during World War I; JDW, Sr. as a child; 1962 Nobel Prize news conference with JDW, Sr., JDW, Jr., and Betty Myers; JDW, Sr. holding his young son JDW, Jr.; father and son in 1949 with the family's new car; JDW, Sr. with his children Jim at age 11 and Betty at age 9. All photos courtesy of the James D. Watson Collection, Cold Spring Harbor Laboratory Archives.

Back cover artwork: (*Top left*) JDW, Sr. in the 1950s; (*bottom left*) JDW, Sr.'s grandfather Delavan Ford with his grandsons Thomas Tolman Watson II, James Dewey Watson, and William Weldon Watson; (*top right*) Jean and James Watson at Indiana Dunes State Park in 1951. All photos courtesy of the James D. Watson Collection, Cold Spring Harbor Laboratory Archives.

Library of Congress Cataloging-in-Publication Data

Watson, James D., 1928-
Father to son: truth, reason, and decency / James D. Watson.
     pages cm
Summary: "Nobelist James D. Watson delves into his family history, exploring his ancestors' roots in Springfield, Illinois, and Chicago, and then focuses on his father James D. Watson, Sr., and his influence on Dr. Watson's success as an eminent scientist and as a writer. Contiguous people, such as Abraham Lincoln and Orson Welles, and events, such as the Leopold and Loeb "Crime of the Century" and 20th century developments in American politics and education, provide a framework for these explorations"--Provided by publisher.
   Includes bibliographical references and index.
   ISBN 978-1-62182-035-2 (alkaline paper)
1.   Watson, James D., 1928–2.   Watson, James Dewey, 1897–1968. 3.   Fathers and sons--United States. 4.   Watson, James D., 1928--Family. 5.   Watson, James Dewey, 1897–1968--Family. 6.   Springfield (Ill.)--Genealogy. 7.   Chicago (Ill.)--Genealogy. 8.   Molecular biologists--United States--Biography. 9.   United States--History--20th century--Biography. I. Title.

   QH31.W34A3 2014
   977.3'56044--dc23

                                    2013049622

10 9 8 7 6 5 4 3 2 1

All World Wide Web addresses are accurate to the best of our knowledge at the time of printing.

For a catalog of all Cold Spring Harbor Laboratory Press publications, visit our website at www.cshlpress.org.

For Rufus, Duncan and Angus

Contents

Introduction

Nobel Prize winners do not come in a standard form. As Master of Clare College, it has been my privilege and a pleasure to get to know two great, but very different, American Nobel Prize winners, honorary fellows of the College. Norman Ramsey came to Clare in the 1930s from Columbia University with little knowledge of physics and worked in the Cavendish with Rutherford. He went back to the United States, to Los Alamos where he liaised with the Air Force on the size of plane needed to drop the atomic bomb. After the war as chair of the Harvard Physics Department he defended younger colleagues against charges of communist sympathies, including a memorable 1953 appearance on *Meet the Press*, which resulted in a dinner with a grudgingly admiring Joe McCarthy himself. Ramsey won the 1989 Nobel Prize in Physics for his work on the atomic clock. He was a progressive—brilliant, courageous, and understated. He did not shirk publicity nor did he seek it out. He was not a household name.

James Dewey Watson, Jr. likewise is a progressive—brilliant and courageous—but bubbles over publicly with new ideas. As I discovered, he can be recognized, out of the blue, by an Italian TV crew in a restaurant on 53rd Street in New York City. He has been an extraordinarily generous benefactor of Clare College, where he was a member when he did the postdoctoral research that led to the discovery of the structure of DNA, celebrated by a sculpture in the grounds of Memorial Court, Clare, and by a plaque outside the Eagle Public House. This memoir of his father, James Dewey Watson, Sr., is crucial to understanding the form of scientist that Jim Watson, Jr. has become.

This affectionate account is, at one level, a lyrical memoir of a father, Jim, Sr., and son, Jim, Jr. —throwing baseballs in the yard, going bird-watching together, traveling the world, living close by in Cambridge, Massachusetts, in his father's widowed later years.

At another level, the story is a chronicle of an archetypal American family from before the Civil War to Vietnam. Here are ambitious settlers making it in the Midwest; successful entrepreneurs who are sometimes victims of financial panics, rescued from financial ruin more than once by strong women; Roosevelt Democrats whose party affiliation was determined by the Depression, which left them for over 15 years without a car; beneficiaries of the war because Jim's mother went on to be a personnel manager for the Chicago Red Cross after being a 7th Ward precinct captain in Chicago Mayor Edward Kelly's Democratic machine; prewar, very modest bungalow homeowners who could just afford after WWII to move to the Indiana Dunes on Lake Michigan.

In this typical American chronicle, famous, if not infamous, names keep appearing. The Watson Saloon, actually a high-class confectioners, in Springfield, Illinois, was where Lincoln and his supporters spent election night in 1860 and where they held the party when he set out for Washington with an American flag sewn by Jim's great, great, great aunt Abigail Watson Ives. Jim's great uncle Dudley Crafts Watson at the Chicago Art Institute mentored and guided his young cousin Orson Welles. Jim's uncle Bill Watson was a Yale professor who worked on the Manhattan Project. Future president of the University of Chicago, Robert Maynard Hutchins, was friend and classmate of Jim's father at Oberlin College. Above all, the precocious, young bird-watching companion of Jim's father in the early 1920s was Nathan Leopold, perpetrator with Richard Loeb of the appalling May 1924 abduction and murder of Bobby Franks that captured the American imagination.

The up and down story of Jim Watson, Sr. also tells a lot about the future Nobel Prize winner. Jim, Sr. had to leave college after a year and spent his life as an accounts manager at LaSalle Extension University. At all times he was a voracious reader and second-hand book collector, much pleased when George Steiner persuaded him to give his collection of novels by John Cowper Powys to Churchill College. He wrote in his diary in 1939, "To me an interesting person is one alive with intellectual curiosity." Intellectual curiosity is something that still consumes Jim, Jr. as his work on cancer testifies.

Jim, Sr. was also impatient with boring people. "There is a kind of optimist who maintains that one should never be bored, that it indicates a defect on the part of the bored person. One ought, the optimist says cheerily, to find interest in anything and everything. This is nonsense. People ought to get bored much quicker than they do, and insist on having something interesting. I would say that the civilized man is distinguished by his capacity for being bored and finding a remedy for it in thought and action." It is perhaps not surprising that his son would write *Avoid Boring People*.

When Jim, Jr. was writing *The Double Helix*, his father counseled him, "In regard to the new book, remember what I said about calling any living person stupid etc. You will find it can't be done." I'll leave others to decide whether Jim heeded that caution.

Jim, Sr. thought hard about political philosophy. In 1957 he worked hard to crystalize his lifetime thoughts in a paper for a small discussion group:

> I would like to talk with you this evening in regard to the philosophy of liberalism. Judging from the criticism directed against this philosophy in recent years, it seems to be defensive. Without exaggeration all of the evils of the present-day world have been placed on its doorstep. And as a result there seems to be a failure of faith in this old traditional philosophy in some quarters.
>
> To those of us who have been liberals for many years it is discouraging to see the ideas with which we have lived discredited. I do not believe that the

liberal ideas are the cause of our troubles, or that this philosophy has outlived its day. To me there seems to be enough human reason left to which to appeal against reckless fanaticism of every sort.

At a time when American politics is wracked by a seemingly remorseless anti-intellectual fundamentalism, the appeal to reason of Watson, father and son, should be cherished.

Tony Badger
Master of Clare College
Paul Mellon Professor of American History
Cambridge University

Foreword

Several years ago, in transferring personal Watson family papers to the Cold Spring Harbor Laboratory Archives, I carefully looked for the first time at the contents of some 30 file folders that became mine after Dad died during the summer of 1968.

Most of Dad's file folders were filled with articles clipped from newspapers and magazines about long-admired writers of literature and liberal politics. Several more personal files, however, caught my serious attention. One contained some 30 letters he wrote back to his family while in 1918–1919 wartime service in France. Another contained descriptions of some 15 bird walks he made to observe the beginnings of the spring migration during the late winter and early spring of 1914. Then he was a junior at Lyon County High School in the Chicago suburb of LaGrange, some 15 miles southwest of Chicago's Loop. Likely also of importance to Dad was an essay exam for his freshman bible class at Oberlin College. What he then had learned likely converted him from the Episcopal faith of his parents to being an agnostic for the rest of his life. Other folders contained diary pages on which he wrote down details of his regular sea journeys abroad after my mother Jean died in the spring of 1957. Particularly revealing are his near end of life impressions of the fellow guests he spent several winters with in a small upper-middle-class resort south of Sarasota, Florida. The genteel world that he had been born into was then not at all what he wanted from life.

In finding that Dad had been a natural, high-quality writer since his high school days, I first considered preparing a small private book containing most of his writings for distribution to Watson family members.

Soon, however, I realized that Dad's documents would become much more useful if I wrote accompanying sections that described my father's activities during successive stages of his life. Originally I saw the need for five such chapters. By now there are nine, each enriched by personal details about key intellectuals he knew through their writings or teachings or through personal contact. The educator, Robert Hutchins, first came into Dad's life when they were both Oberlin College freshmen.

Most rewarding for me to put together has been Chapter 1, which I start in Springfield, Illinois. My own branch of Watsons there put down their first solid Midwestern roots. The homes of William Weldon Watson and his son, Benjamin, were located in close proximity to Abraham Lincoln's for more than 20 years. The first Springfield-rooted Watsons were exceptional frontier entrepreneurs who took great risks to give themselves and their families' children the opportunity to aim for the top of late-19th-century Midwestern life. To give his and his father's "Watson & Son Confectionery" sufficient capital to grow, Benjamin A. Watson saw no alternative but to join the California Gold Rush of 1848–1850. In so doing, he left behind his pregnant wife Emily and their year-old son William Weldon Watson III. I would not now be putting together this book if he had not come back sufficiently cash-flush to also let him subsequently build the Midwest's first post–Civil War truly grand mineral springs resort nestled in Perry Springs near the Illinois River between Springfield and St. Louis. Nor would his first son, William Weldon Watson III, have opened on Dearborn Street Chicago's first French restaurant in 1882. Already by then WWWIII was proprietor of the Whiting House, the then-great resort on the shores of Wisconsin's Lake Geneva, just across the Illinois border and then-preferred site for wealthy Chicago residents to escape the blistering heat that dominated most of Chicago summers. WWWIII with his wife and six children lived outside of town on a farm that let them continue to think of themselves as part of the frontier.

Dad's later more cautious approach to risk taking is easily explained by his having no desire to repeat his father's disastrous failure to find gold on the penny stock market. Equally important may have been the calming influence of his conservative, churchgoing mother, Nellie Dewey Ford, whose grandparents' oil portraits have long dominated our Cold Spring Harbor home's front hallway. Their stern faces suggest that life was rough and tough for those new frontiersmen and -women carving out permanent existences on the lands where prior Indianan bands never themselves had ever seen fit to put down their own roots.

Acknowledgments

Father to Son, in its transition from manuscript to final book, contains more social history than initially envisioned because of the diversely informative sidebars assembled by Judy Cuddihy in her Developmental Editor role. Her inquisitive, highly intelligent mind has vastly improved virtually every page of this book. Through creating appropriate pull-offs of my father's Winter/Florida diary entries, Judy gets to the heart of who he was: for example, why, when not with his family, he invariably chose books over our neighbors. Judy would have much enjoyed conversations with Dad if a non-cigarette way of life had let him live long enough to witness the early days of the Cold Spring Harbor Laboratory Press, which had the good fortune to have had Judy on its staff since 1979.

The "going for gold" theme of Chapter 1 came about through Mary Coe Foran, a direct descendent of William Weldon Watson I, writing me of the existence of WWWI's son Benjamin's letters to his wife in the California Gold Rush days of 1849–1850 and their current availability through the Gold Rush Museum of Sacramento, California. Bernard Ewell, the noted appraiser of Salvador Dali's artwork and the grandson of my father's Uncle Dudley Crafts Watson, first let me know his great grandfather WWWIII's involvement in the Honduran search for gold at the end of the 19th century.

CHAPTER 1

Going for Gold

My father, James Dewey Watson (b. 1897), through both his father, Thomas Tolman Watson (b. 1876), and his mother, Nellie Dewey Ford (b. 1875), descended from ambitious, hard-working settlers who moved west to seek more fulfilling lives as Midwesterners than they would have had they remained at the Eastern sites of their births. Nellie Dewey Ford's maternal grandparents, James Johnson Dewey (b. 1814) and Eliza Ann Bates (b. 1816), moved directly to southern Wisconsin, where they helped found the town of Geneva (later changed to Lake Geneva, just to the north of its border with Illinois). They came to the Midwest from Cooperstown, where several generations of Deweys had resided, all descendants of Thomas Dewey, "the settler," who arrived in Boston from England in 1630. In Lake Geneva James Dewey became its first hat merchant, going on to serve for several years as Lake Geneva's postmaster and then its village president in 1851. In the early 1870s, James and Eliza's daughter, Sarah Loomis Dewey (b. 1839 Cooperstown), married Nellie's father, the prosperous lumber dealer, Delavan Ford (b. 1825 or 1830/1832), whose parents, Jonathan H. Ford (b. 1794) and Mary "Polly" Jessup (b. 1800), likewise were born in New York State.

The route of Dad's earliest known Watson ancestor, William Weldon Watson (WWWI, 1794–1874), to the newly established state capital of Springfield, Illinois, was less direct. Born in the town of Sussex, nestled in the low mountains of northwestern New Jersey, his own ancestors were likely among its early settlers as witnessed by the still-existing names of Weldon Road and Weldon Brook. Early in life, WWWI became a silversmith, beginning to learn his trade in Morristown and mastering

James Johnson Dewey (1814–1898), Nellie Dewey Ford's grandfather.

Eliza Bates Dewey (1816–1858), Nellie Dewey Ford's grandmother.

it quickly so that before his majority he was in charge of a large number of unskilled workers at a Newark, New Jersey, candle factory.

In moving west across the Appalachian Mountains in 1817, WWWI left behind in Morristown three sisters, all of whom also eventually married and lived into their eighties. In Lexington, Kentucky, he had the good fortune to meet and soon marry the very young widow Maria Cape Humerickhouse. Why WWWI and Maria decided afterward to quickly move south and west to Nashville, Tennessee, we will likely never know. Their first child and only son, Benjamin A. Watson, was born soon after they arrived on December 9, 1818, and commenced to live on a farm several miles from the city center.

WWWI, besides continuing to ply his silversmith trade, also functioned as the minister of an aspiring Baptist church, whose sermons he spiritedly delivered for the next 15 years. To combat the rampant alcoholism of early settler life, WWWI, upon returning from a Baptist gathering in Philadelphia, brought with him by prairie schooner the first soda water fountain ever seen in Tennessee. Setting it upon the street corner of his church, he reputedly made soda water the rage of the Nashville of his time. According to family lore, he quickly made enough money to build a new church, which still stands in the heart of Nashville. Before the premature death of Maria in 1834, she had borne him four daughters beside Ben—Abigail in November 1822, Ann Marie in December 1824, Hester in July 1826, and Cordelia in March 1828.

Perhaps then overwhelmed by the need for a better-paying occupation to care for his four motherless daughters, WWWI moved to St. Louis for two years to learn the confectionary business. From there he and his five children moved back across the Mississippi to Springfield, Illinois, just before the Illinois legislature, led by its Long Nine, was about to vote to move their state capital there. There WWWI opened a confectionary shop, Springfield's first, using skills acquired during his previous time in St. Louis. Also then moving to Springfield in response

Abigail Watson's calling card. Eldest daughter of WWWI, she married John G. Ives in 1843.

to the impending arrival of the state legislature was young, 28-year-old attorney Abraham Lincoln. Springfield's central location made it a much more appropriate hometown than his going-nowhere, underpopulated New Salem.

The 1836 Springfield, Illinois, on which Abraham Lincoln and William Weldon Watson then bet their futures was still very much a frontier village, which had been settled around 1820 as "Calhoun" and renamed in 1832. The Indian Wars had subsided and the town's major problems were instead knee-deep mud and hogs roaming at will. In an 1839 newspaper report, Springfield was described as "a throng of stores, taverns, and shops … and an agreeable assemblage of dwelling houses." At first, Springfield was a strong Whig town in a Democratic area. Lincoln himself was a Whig and had served four terms in the State Legislature. But the arrival of a new wave of immigrants in the 1840s shifted the political balance toward the Democrats.[1]

The center of all major activity in Springfield was the public square. Lincoln lived above Joshua Speed's store on the public square when he first came to Springfield, soon joining the law practice of John Todd Stuart in 1837. After a term in the U.S. Congress, he practiced with

[1] **Moving the state capital was driven by the pattern of immigration and the need for better transportation, including access to the railroads being built. The state legislators also said that they were tired of eating prairie chicken and venison in the then capital city Vandalia.**

Stephen T. Logan in 1841, becoming in 1844 a senior partner in a firm with William Herndon as his junior partner. They practiced law at several successive locations on or near the public square, arguing their cases in the courts on the east side of the square, only a few minutes walk from the Watson & Son confectionary saloon on its south side.

On June 27, 1843, WWWI's eldest daughter, Abigail, at age 21 married the highly successful jeweler John G. Ives,[2] who first settled in Springfield in 1839, after coming from Oneida County, New York, where he was born in 1819. Later, as a grain merchant, he was to become one of Springfield's most prosperous citizens, erecting the Aetna Flour Mills in 1855, which he ran for the next 10 years. Abraham Lincoln habitually doffed his stovepipe hat when he came upon John Ives on his way to his own office, also on Market Square. Upon Lincoln's election to the Presidency, Abigail stitched together by hand a large American flag, with its 31 stars painted by her husband. It draped the engine on the train from which Lincoln made his farewell speech before departing to Washington.

That Benjamin would ever have a career that would already at age 30 let him match his eldest sister's prosperous gentility, however, was far from foreordained. Less than a year after he came to Springfield in 1836, he was the defendant in a court action argued by Abraham Lincoln, who successfully obtained a judgment against young Benjamin, then only 19 years of age, of $87.75 payable to the plaintiff with $10.00 payable to Lincoln and his partner, John T. Stuart. Although unable to soon pay off the debt, Ben saw no reason not to continuously expand his horizons. When only 22, he joined nine other local youths in August of 1840 to put a roof on an old stagecoach, powered it by four horses, and drove back and forth over five weeks to Nashville, Tennessee, to hear the great early 19th century Kentucky statesman, orator, and opponent of slavery Henry Clay, then campaigning to be the Whig candidate for the Presidency. They camped out along the way, did their own cooking, and sang campaign songs as they passed through the towns and villages along their routes. In

[2] Ives then served as Treasurer of Sangamon County and Secretary of the Illinois Board of Trade. His and Abigail's son Benjamin was the father of Orson Welles's mother, Beatrice Ives (1882–1924).

Cordials, Candies, &c.

THE subscriber having commenced the Confectionary business in Springfield, in the house recently erected by Dr. Gray, on the corner of fourth and main streets is now prepared to furnish any article in that line on as good terms as can be bought in St. Louis, or Alton, and warranted not to be inferior to any manufactured in the Western country.

The ladies of Springfield are respectfully informed that they can be furnished with all kinds of Cakes and Pastry on the shortest notice, made in the best manner and dressed in the neatest style.

A constant supply of Cakes, etc. always on hand.

W. W. WATSON.

Springfield, sept. 15, 1836.   b54

Advertisement announcing the opening of Watson's Confectionary in September of 1836. Throughout the years, this establishment sold a broad range of products, including soda crackers "expressly designed for dyspeptics," fruit cake and plain cake, Watson's cough candy, leather preservative, waterproof oil blacking, guitar strings, fishing tackle, and musical instruments.

The W.W. Watson & Son confectionary establishment was located on the south side of the Springfield square apparently next to T.S. Little's clothing store. This photograph was taken by Preston Butler around 1859, and his studio is also shown here. The Statehouse on the square was the site of Lincoln's funeral in 1865.

some they were applauded, in others jeered, and occasionally they were pelted with eggs. In Nashville, where 40,000 people were assembled, they were invited onto the stage to sing their spirited tunes. Afterward "they felt emotionally well paid for their time, labor and expense."[3]

With his father WWWI now remarried for the past three years, Benjamin, in turn, married Emily R. Planck on February 11, 1845. The first of their seven living children, whom they named after his father, was born on May 13, 1847. Called "Welly" in person, my father's grandfather bore most of his life the WWWIII designation.[4] Whether WWWI

[3] In 1839, W.W. Watson advertised the location of the Confectionary at the corner of Main Street north of Hoffman's row. On October 21st of the same year, he announced that his son B.A. Watson was being taken into the business as co-partner and "business will in the future be carried on under the firm of W.W. Watson & Son."

[4] Aside from WWWIII, Benjamin's other sons did not fare well. His namesake Benjamin Jr. died of dysentery before his first birthday, and Bertram succumbed to scarlet fever at age two and one-half. An older son, Harry, died in a railroad accident at age 18 years.

SCHEDULE I.—Free Inhabitants in *Springfield* in the County of *Sangamon* **State** of *Illinois* enumerated by me, on the *8th* day of *November* 1850. *S. B. Moody,* **Ass't Marshal.**

123

Dwelling-houses numbered in the order of visitation.	Families numbered in the order of visitation.	The Name of every Person whose usual place of abode on the first day of June, 1850, was in this family.	Age.	Sex.	White, black, or mulatto.	Profession, Occupation, or Trade of each Male Person over 15 years of age.	Value of Real Estate owned.	PLACE OF BIRTH. Naming the State, Territory, or Country.	Married within the year.	Attended School within the year.	Persons over 20 y'rs of age who cannot read & write.	Whether deaf and dumb, blind, insane, idiotic, pauper, or convict.	
1	2	3	4	5	6	7	8	9	10	11	12	13	
		Catharine Gough	½	F				Ills					1
		Mary Murphy	24	F				Ireland					2
34	757	Wm. W. Watson	52	M		Confectioner	860	N.J.					3
		Sarah Watson	47	F				N.Y.					4
		Emily Watson	21	F				Ills					5
		Wm. W. Watson	3	M				"					6
		Emily Watson	1	F				"					7
		James King	16	M		Confectioner		N.J.					8
		Herman Bodecker	38	M		Clerk		Germany					9
5	758	John C. Mavey	36	M		Livery Stable Keeper		Ky					10
		F. J. Mavey	36	F				"					11
		Margaret Mavey	13	F				Ills		1			12
		James Mavey	11	M				"		1			13
		Mary A. Mavey	9	F				"		1			14
		Maria Mavey	3	F				"					15
		Zachariah Mavey	10/12	M				"					16
		Patrick Molony	19	M		Laborer		Ireland					17
		John Brady	24	M				"					18
		Edward Fouring	13	M				"					19
		Sarah Millbeck	20	F				Ills					20
		Ellen Shrow	20	F				England					21
6	759	Cornelius Groesbeck	33	M		Laborer		N.Y.					22
		Rebecca Groesbeck	34	F				"					23
		Harriet Groesbeck	10	F				Ills		1			24
		Allia J. Groesbeck	2	F				"					25
7	760	Enoch Orahood	22	M		Waggon maker		Ohio					26
		Ann Orahood	21	F				Pa					27
		Ed. D. Orahood	½	M				Ills					28
	761	B. C. Pabon	40	M		Carpenter	750	Ohio					29

1850 United States Census showing the WWWI household in Springfield. William Weldon Watson III, age 3, is listed here, as is Benjamin's wife, Emily.

ever fathered a very short-lived WWWII will likely never be known. Despairing of ever having the financial means to give his own children as well as his father secure futures, Ben saw the finding of gold in California as the way to end his financial uncertainty and quickly incorporated a company of 10 investors to let him lead a group of men across the western plains to the steep western slopes of the California Sierra Mountains where the first big gold strike took place.[5] Although he never found much gold by himself, he was soon earning $700 per month providing supplies to the actual miners. Away for less than two years, he returned to Springfield modestly rich. Only a fraction of his new money was needed to let his and his father's store soon become Springfield's fanciest confectionery store, Watson Saloon.[6]

[5] **News of the discovery of gold in California on public lands had reached the eastern United States in mid-1848. Benjamin arrived in California around September of 1849 and first worked in the mines in Lassens, the Reddings Diggings, and the American River before going to Sacramento.**

[6] **B.A. Watson arrived back in Springfield on January 28, 1851, sailing by way of San Juan de Nicaragua and Chagres on the isthmus of Panama with 500 Californians.**

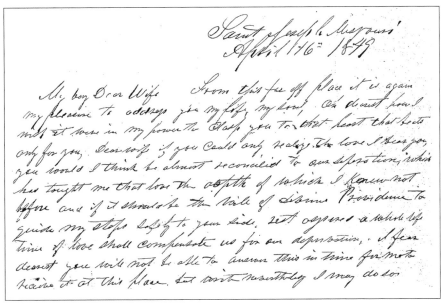

Holograph of one of Benjamin's letters to Emily. From St. Joseph, Missouri, Ben's party traveled through the Nebraska Territory and then took the "Nevada route" to California, where they arrived in August or September of 1849. By November of 1849, the Illinois and California Mutual Insurance Company had been dissolved mainly owing to lack of provisions. Ben spent the winter in Sacramento.

Benjamin Watson left for the California Gold Rush in March of 1849. Here is the address for his wife, Emily, in Springfield.

Camp on the Sacramento
at Lassens, California Sept. 12, 1849

My Dear Wife:

At length I have the inexpressible gratification of addressing you from this long wished for country, from the promised land from the El Dorado of our hopes. We arrived here on Monday at 2 P.M. after enduring almost incredible hardship. We are all well we got all of our wagons in and all of our mules except one which was bitten by a snake and died on the Humboldt River. Lassens is situated on the Sacramento River somewhere what I suppose is the head of the lower valley, you will perhaps be surprised to find us coming into the country at this point I will tell you how it come to pass about seventy miles above the sink of the [???] River we come up with a man by the name of McGhee who said he was looking for a road or old trail which led off to the Northwest and which he said was a better route and a shorter one. Many of our men tired with the monotonous scenery of the Humboldt River and tired of the bad water and bad grass eagerly caught at the idea. In six miles farther we found the road, it being Saturday Evening we lay over until Monday morning then the company decided upon taking the new road. I [refused?] it because I did not like to try experiments which might prove disastrous. We started on Monday morning and traveled 1 day and night for fifty five miles without grass or water over a plain of burning[?] sand but the sand was fortunately good.

We reached the Black Rock Boiling Spring at 3 P.M. on Tuesday. The spring is boiling hot, one hundred fifty feet in diameter. It nourishes a patch of grass which we found green and good, but this nor the three other hot springs in the next five miles will furnish a sufficient quantity of grass for many teams if they should come this road. Next morning we travelled on to the last of the hot springs where we found very good grass which we cut to take along with us as our next drive was across a sandy salt plain of twenty five miles without grass or water. We started on Thursday morning at 2 A.M. and reached the entrance of High Rock Canon at 5 P.M. after a day of fatigue and toil for men and mules where we found good grass and water. That night the Indians shot a mule and a horse in fifty yards of our guard. They belonged to a pack train camped near us. Our road lay next day over a gradual hill covered with volcanic stone making very hard travelling. We had to let our wagons down with ropes a very steep hill two hundred feet high into a canon or as it is pronounced canion, which is a gorge or narrow valley in the mountains of the foot hills of the Sierra Nevada. We travelled two days up this beautiful pass at times the pass not being more than 30 yards wide and others reaching 200 with perpendicular walls of granite reaching several hundred feet high. Many times on both sides at once our road still continued pretty good, except some rough stones, bearing North and West all the time mostly north.

On Wednesday the 22nd of August we reached the main chasm of the Nevada Mountains and ascended for one hour when we encamped near a brook of good water under some fine pine trees being the first we have camped under trees since we left the vicinity of the Missouri River. Next morning we doubled teams and by 12 o'clock we had all of our wagons on the top of the Sierra Nevada but it was accomplished by immense labor of men and mules we drove down into a nice little valley and encamped for the day. Our camp is surrounded by mountains covered with majestic pine trees. Next day we struck [???] Lake or Pitt Lake when the road turned south road good grass on the lake very good. Next day struck Pitt River which is the East branch of the Sacramento. Not knowing the name of it I had called it Pyramid Creek from a number of stoney formations standing in a cluster and resembling pyramids in shape. We traveled down this several days the road frequently diverging from the river to avoid canons and [???] in the river. On Thursday 30th Aug. At noon, we met a train of packers from California for Oregon who had been down in the mines digging gold. They showed us lots of the yellow stuff they told us we were 180 miles yet from Lassens. Our road now became as bad as a road could be and be passable for wagons. By reference to Fremont's map you will find Round Valley. Then strike a Southwest course to the mouth of Deer Creek on the Sacramento and you will see our road lay through

The Illinois and California Mining Mutual Insurance Company, Leave this city, for their destination, at two o'clock to-day. They seem to be well prepared for their arduous undertaking.

Our townsmen will see in this list many of our old city residents and valuable citizens. Others are from the county. Among them are farmers, mechanics and professional men. Our best wishes go with them. May they realize their hopes and return in safety to their families and friends.

There are two other companies formed here, which will leave soon.

NAMES OF THE MEMBERS.

B. A. Watson,	T. Billson,
C. E. White,	Lewis Johnson,
Albert Sattley,	John Rodham,
Benj. F. Taylor,	Richard Hodge,
E. Fuller,	Jacob Uhler,
Wm. B. Broadwell,	E. R. Biddle,
W. P. Smith,	J. B. Weber,
B. D. Reeves,	John B. Watson,
Wm. Odenheimer,	F. S. Dean,
Henry Dorand,	Thomas J. Whitehurst.
E. T. Cabaniss,	

Newspaper notice of B.A. Watson's departure for the California gold fields. In a February 24, 1850, letter, B.A. Watson writes that "there is gold upon every square yard of the country for miles around—but it requires work! work! work! to find it." He also relates that many of the proceeds are lost at the gaming tables: "Nearly everyone gambles. I have seen fortunes won and lost in an hour."

B.A. Watson's miner's leather pouch and a quartz/gold nugget weighing 1.28 troy ounces.

(Continued from previous page.)

the mountains the whole way some of the road was pretty good all of it except the last fifty miles lay through an immense pine forest of the most magnificent trees I ever beheld some of them measuring 25 feet in circumference and 200 feet high and 80 feet without a limb and perfectly straight. The last forty miles has been one of constant toil and hardship. The road running down one of the spines[?] of the mountains. We capsized one of [the] wagons and smashed the top in, broke a number of the wheels of the others but which we temporarily repaired and was three days in traveling 30 miles never finding water in less than 1/2 mile of the road and that all the way down some steep and rugged mountain side. We found but very little grass for our mules.

About fifty miles from this place we met an expedition of the government going out to explore a railroad route for the Humboldt River an expedition some 80 men strong at an expense of 18 or 20,000 dollars per month. They giving all their men from $180 to $1,000 per month truly no man knows the wastefulness of our government. We are now laying here repairing our wagons two of which will start to- morrow for Sacramento City for provisions. Three men will go down with them the balance of us will start at the same time for the upper diggings 60 miles North of us (there is a man now here who has been working in the mines and who showed us a plate full of gold, all larger pieces he says we can make $75.00 per man per day). Dear if so I will soon have as much as we will want and then away for home and its sweet enjoyments. I fear it will be some six weeks before I shall be able to hear from you and we have no post office in less than two hundred miles of us but we are going to make an effort to get one in our imme- diate vicinity though I may be more fortunate than the balance as I directed Cook Matheny who had with R. McDaniel left Eli Cook and was going through on pack mules to write to the post mission at San Francisco to forward my letters to Sacramento City. if he attends to it I will receive my letters by the return of our wagons in three weeks. I ex- pect to go down myself when the wagons return. I think the prospects are full as good as if not better than when I left home. I feel almost *confident of making a fortune in the coming year and if so hey day for a merry time we will have and that brings me to another idea. If our expectations were well formed you should about this time be giving me another pledge of love of that if so I need tell you cheer up and be of good heart for I know you will be so with your beautiful children around you and the thought that while you are reading this your husband is accumulating gold enough to make you and them comfortable and independent for life and that already that is when you receive this one third of the time will have already expired for which he left you and them. Have no fear of our suffering for provisions as accounts from below show that there is more in San Francisco than they can possibly consume in the country in the next twelve months and more arriving every day. As for personal security we are assured by those have spent some time in the mines that a man's more safe there than he would be in St. Louis. The miners have established a code of laws for their own protection which have been found amply sufficient. I must now bid you farewell again my love. I shall endeavor to keep you regularly informed of our success. There will be a mail once a week from the [???] for the states. You must direct your letters to Sacramento City California. Tell Hetty to do the same with directions to the post master to keep them till called for.*

Tell my noble boy my Weldon, my first born to be a good boy and father will bring him a pony when he comes back, and lots of pretty things for the baby. Tell Father to write to me and tell me the news in general about business and tell Walter and John to write also. Take good care of your health. Send me a dozen kisses in you next letter. I must now bring my letter to a close. I should like to give you a description of the natives of this country but I must defer it to some future time. Suffice to say Fremont's description is correct about their wardrobe a bunch of grass serves the females for a fig leaf, the [???] of [??? ???] ever so slight a covering. Farewell my adorable lovely beautiful affectionate wife. Believe me more than ever your affectionate husband.

B.A. Watson

Benjamin's signature on his copy of the Constitution of the Illinois and California Mutual Insurance Company, No. 1. Members going to California gold fields had to pay $150 to the association before leaving. Once everyone returned to Springfield, the assets of the association were to be divided equally among the 21 signatories. "The labors of this association in California shall cease at the beginning of the rainy season in the fall of 1850." The members mutually pledged "to each other, our lives, our fortunes, and our sacred honor."

CONSTITUTION

OF THE

ILLINOIS AND CALIFORNIA MUTUAL INSURANCE COMPANY, No. 1.

Adopted on the 20th of January, 1849, in the city of Springfield, Illinois.

Section 1. This association shall be called THE ILLINOIS AND CALIFORNIA MUTUAL INSURANCE COMPANY, No. 1.

§ 2. Its officers shall consist of a President, Secretary, Treasurer and Superintendant, who shall be elected on the first Monday of every month.

§ 3. It shall be the duty of the President to preside on all occasions when the Association may be in session; or during his inability to act, the chair shall be filled by a vote of the association.

§ 4. The Secretary shall keep an account of our travels, together with a correct record of the acts of the association, and issue orders upon the Treasurer for all sums appropriated, which orders shall be countersigned by the President.

§ 5. The Treasurer shall receive all monies belonging to the association, none of which shall be paid out, or in any way used except in payment of orders from the Secretary, countersigned by the President, and in accordance with appropriations made by the association.

§ 6. The Superintendant shall on all occasions, superintend the operations of the association.

§ 7. Standing, or special committees, may be appointed at any time, to render services not otherwise provided for; such committes to be appointed by the President, unless the association otherwise direct.

§ 8. No person shall become a member of this association without the consent of all its members.

§ 9. Every member shall pay into the treasury, on or before the 10th day of February next, the sum of one hundred dollars; and on or before the day set apart for leaving Springfield, the sum of fifty dollars; such payment when made to become the property and funds of the association.

§ 10. Any member failing to pay the first instalment, as provided by the 9th section, shall cease to be a member, and any member failing to pay the second instalment, shall cease to be a member and forfeit his first instalment.

§ 11. No member shall withdraw from the association without the consent of all its members; but a member may be expelled by a vote of three-fourths of its members, with the understanding that such member shall be entitled to receive an equal share of the nett profits or income of the association up to the time of his expulsion, after deducting the expenses and charges therefrom; and provided that the outfit, property and materials of every kind used by the said association in the course of its business for the prosecution thereof, shall not be considered profits or income subject to such division.

§ 12. Members shall not be allowed to quarrel among themselves, nor shall any member be allowed to drink intoxicating liquors, gamble, use profane language or labor on the Sabbath.

§ 13. All discussions between members of this association of a character tending to create discord and contentions shall be carefully avoided.

§ 14. During the term for which we are connected with any member may devote his time to that kind of employment supposed to be most profitable—provided three-fourths of its members concur therein—but the proceeds of his labor must be deposited with the Treasurer of the association.

§ 15. All property, money, estate, claim, or thing, or right, having value, which may be acquired by the association, or any member thereof, shall belong to the association in its aggregate capacity; nor shall any member in his individual capacity, unless permitted to do so, be entitled to claim or receive any part thereof to his separate use, except as provided in the 17th section.

§ 16. The labors of this association in California shall cease at the beginning of the rainy season in the fall of 1850.

§ 17. On the return of the association to the city of Springfield, Illinois, a division of all its assets shall be made among its members—each having an equal share, without deduction or abatement, and the Treasurer shall, on the drafts of the Secretary, countersigned by the President, pay the same: *Provided*, That if any member shall at any time after the association shall have ceased its labors in California, for the purpose of returning home, desire to withdraw from the association, he may do so, and shall be entitled to receive at such time his share, as specified in this section: *and provided further*, that special appropriations may be made from time to time, to individual members for their support and private use, of which correct account shall be kept, and said amount deducted from the the amount or shares due to such members severally, on the final division of the assets of the association.

§ 18. In case of the sickness or death of any person while a member, after leaving Springfield, Illinois, his share shall be set apart as if he were living, and deposited with the Treasurer of the State of Illinois for safe keeping, subject to the order of the legal representatives of said member, to be disposed of according to law.

§ 19. The association shall terminate its existence and be dissolved as soon after its return to the city of Springfield, Illinois, as said distribution of its assets can be made, and a final settlement made of its affairs.

§ 20. All disputes, claims or questions, arising between the members of the association, or between the association and any of its members, respecting their rights or obligations under these articles, shall be determined by the vote of the association, and such decision shall be final and conclusive; and any member who shall apply to any court or other jurisdiction whatever, previous to the return of the association to the city of Springfield, Illinois, shall forfeit all his rights as a member of the association, and all claims to any share of its assets.

§ 21. Two-thirds of its members shall constitute a quorem to do business.

§ 22. These laws may be altered or amended, by a vote of three-fourths of the members, except sections 14, 15, 17, 18 and 19, which shall remain unalterable.

§ 23. The foregoing provisions shall be strictly adhered to as far as practicable; in support of which, with a firm reliance on Divine protection, we mutually pledge to each other, OUR LIVES, OUR FORTUNES, AND OUR SACRED HONOR.

Wm. C. Broadwell   Henry Dorand   Jacob Uhler

B. A. Watson   E. S. Cabanis   B. R. Biddle

N. P. Smith   T. Billson   J. Weber

O. E. White   Lewis Johnson   John W. Watson

Wm. Osenheimer   H. S. Reeves   F. S. Dean

E. Fuller   John Redhaw   Albert Vattley

P. Heslep   Benj. F. Taylor   Thos. J. Whitehurst

Family lore passed on to my father from Welly's son Dudley Crafts incorrectly reported that his grandfather Benjamin was the architect of Abraham Lincoln's home. Originally designed and built in 1839 as a one-and-a-half story home for the Reverend Mr. Dresser, a local minister, it was not lived in by the Lincolns until after their 1844 marriage. The house on Washington Street that WWWI bought after his remarriage in 1842 to the widow Sarah Wiley Mottashed was only four blocks away from the house the Lincolns would later buy. His son Benjamin's eventual home on Monroe Street, between Eighth and Ninth Streets, was located even closer—only two blocks from the Lincoln home on the corner of Jackson Street and Eighth Street.[7] Only some 10 years later, in the middle 1850s, did Mary Todd Lincoln arrange for it to be raised to a full two-sto-

[7] **Physician Preston H. Bailhache described Springfield in 1857 as "more of a village than a city, where everybody knew everybody." There was a "feeling of equality... so that Mr. Lincoln's appearance on the streets did not at first seem to differ from the other men we met."**

A May 20, 1865, Harpers Weekly *depiction of Lincoln's Springfield house.*

Map of the central area of Springfield, Illinois, drawn by A. Ruger in 1867. Lincoln's house is at the corner of Jackson Street and Eighth Street, B.A. Watson's house was located on Monroe between Eighth and Ninth, and WWWI's home was on Washington Street between Seventh and Eighth.

[8] Caroline Owsley Brown of Springfield related that "Mr. Watson, the confectioner, way back in the forties, had gone to St. Louis to learn to make the famous macaroon pyramids," often referred to as "these sticky monuments."

ry residence. Whether Benjamin ever helped in the addition's design we shall never know. We can be sure, however, that by the mid-1850s the Lincolns were, in fact, major patrons of W.W. Watson & Son confectioners, using them to provide the fine delicacies that flowed in abundance as Mrs. Lincoln furthered her husband's legal and political careers. Mary Lincoln hosted large receptions to which as many as 500 persons were invited. Featured at all such parties were the Watson's famous macaroon pyramids, without which no party table was complete.[8]

In late 1855, Abraham Lincoln took out $107.36 from his account at the Springfield Marine and Fire Insurance Company and wrote a check for $41.72 to W.W. Watson & Son to pay for holiday parties. Five years later, late on the night the Presidency would be decided, Abraham Lincoln came to Watson's Saloon (no alcohol, largely candies and ice cream), which the Republican women of Springfield (now 10,000 inhabitants) had taken over for the night. Mrs. James C. Conkling of Springfield related that "The supper at Watson that night was a rich and very exciting scene," and the *Missouri Democrat* newspaper reported that Lincoln "came near to be killed by kindness as a man can without serious results" at Watson's.[9] Four months later, on the night before Lincoln took off with his inaugural party to Washington, he was again at Watson's enjoying fresh Baltimore oysters in cans, peaches and pineapple, cheese and sardines, almonds, and fireworks.[10]

Once the Civil War commenced, concerns among Springfield Republican Party members became increasingly expressed over lucrative financial war contracts from the Quartermaster's and Commissary departments going preferentially to Democrats. In May 1863, Benjamin

[9] Lincoln waited at Watson's for the election returns from New York, which finally gave him the victory. But Lincoln carried Springfield by only 69 votes against his principal opponent Stephen A. Douglas. Dr. Preston H. Bailhache reported that "Springfield was the mecca of central Illinois on that night of all nights when lightning flashed over the wires that startling news that Abraham Lincoln was elected President of the United States." For those gathered at "Ben Watson's big Ice Cream Saloon … campaign songs were sung and gaiety was the order of the night." "Bedlam let loose might describe it." When Mr. Lincoln slipped out quietly, he looked grave and anxious.

A. Lincoln's December 19, 1855, check to W.W. Watson & Son. The State Legislature met during the first part of the year; thus, the Lincolns did extensive entertainment during the Christmas and New Year holidays.

[10] There are several versions of Lincoln's speech on leaving Springfield, but in all he pays tribute to his years in Springfield. "My friends—No one, not in my situation, can appreciate my feeling of sadness at this parting. To this place, and the kindness of these people, I owe everything. Here I have lived a quarter of a century, and have passed from a young to an old man."

[11] In May of 1860, T.S. Little's clothing store sold its stock to W.W. Watson & Son. Watson advertised: "Having purchased the stock of clothing of T.S. Little, we are now ready to furnish all who are in want of any thing in the clothing line, at greatly reduced prices, FOR CASH ONLY."

[12] In an 1864 endorsement on a letter from B.A. Watson requesting a commission for Post Sutler for Camp Butler, Lincoln wrote that B.A. Watson was "an intimate acquaintance and friend, and is of good character."

Watson became co-signee to a letter written by the prominent Springfield Republican leader Jesse K. Dubow, who objected to the appointment as Commissary of Ninian Edwards, Lincoln's brother-in-law and long a partisan Democratic supporter of Stephen A. Douglas, and of William H. Bailhache as Quartermaster General. Their letter quickly had its desired effect. Both Edwards and Bailhache were soon removed from Springfield, but at the same time allowed to keep their commissions.[11,12]

In this letter Lincoln lists B.A. Watson among others to be considered for the posts of Quartermaster and Army Commissary to replace Edwards and Bailhache. The group met on May 29, 1863, and recommended the appointment of George R. Weber as Commissary and James Campbell as Quartermaster, to which Lincoln agreed on June 22, 1863.

After the Civil War had ended and the tragic death of Abraham Lincoln left a sad, vacant mood over Springfield's citizens, Benjamin Watson and his family moved 60 miles to the west to Pike County. There he and fellow Springfield investors were soon to create a new grand railroad-connected summer resort amidst the health-promoting outpourings of mineral springs. Already on the site were a much more modest hotel and surrounding cabins making use of what the remaining local

A view of the extensive buildings and grounds of the Perry Springs Hotel that appeared in the 1872 Pike County Atlas.

PRIVATE LAWS

OF THE

STATE OF ILLINOIS,

PASSED BY THE

TWENTY-FIFTH GENERAL ASSEMBLY,

CONVENED JANUARY 7, 1867.

VOLUME II.

SPRINGFIELD:
BAKER, BAILHACHE & CO., PRINTERS.

1867.

AN ACT to incorporate the Perry Springs Hotel and Railroad Company.

In force March 7, 1867.

SECTION 1. *Be it enacted by the People of the State of Illinois, represented in the General Assembly,* That B. A. Watson, Noah Divelbiss, O. A. Savage, R. B. Hatch, C. L. Goltra, H. I. Noyes and W. W. Watson, of Illinois, and Dwight Durkee, E. Williams Fox and John H. Seagrist, of St. Louis, and their associates and successors, are hereby created a body corporate and politic, by the name and style of the "Perry Springs Hotel and Railroad Company," with perpetual succession, having power to sue and be sued, plead and be impleaded, in all courts, either in law or equity; to use a corporate seal, issue stock bonds and other securities and evidence of indebtedness, and to receive the same; to have a capital of two hundred thousand dollars, which may be increased to five hundred thousand dollars, to be divided into shares of one hundred dollars each, and to do all things necessary for the purposes of said corporation. *Corporators. Name and style. Powers. Capital stock.*

§ 2. Said corporation shall have the power to maintain the hotel now in operation at Perry Springs, Pike county, state of Illinois, and to erect such other buildings as the business of said corporation may require, and to buy and sell such additional real estate as may be required for the use of said corporation. *Business and operations.*

§ 3. Said corporation shall have the power to locate, build and operate a railroad, commencing at some point on the Toledo, Wabash and Western Railroad, between Meredosia and Versailles, passing as near to Perry Springs as may be practicable, and connecting with the line of what is known as the Pike County Railroad, at some point between Salem and Griggsville Landing. The right of way for said railroad to be obtained as provided for in an act to amend the law condemning the right of way, for purposes of internal improvement, approved June 22, 1852. *Railway privileges.*

§ 4. That a meeting of the shareholders shall be held at Perry Springs on the first Monday in April, after the passage of this act, when a president and board of directors may be chosen—each shareholder being entitled to one vote for each share of stock held by him or her, who shall make all necessary rules for said corporation; shall call meetings, elect a secretary and other officers, and do all things requisite to carrying out the objects of this company. *Meeting of shareholders —vote of.*

§ 5. The stock of this company can only be transferred upon the books of the company at their office at Perry Springs. *Transfer stock.*

§ 6. This act shall be deemed a public act, and be in force from and after its passage, and shall be liberally construed in all courts and places for the benefit of the objects herein contemplated. *Construction of act.*

APPROVED March 7, 1867.

Incorporation of the Perry Springs Hotel and Railroad Company on March 7, 1867. Among the incorporators are B.A. Watson and W.W. Watson. The corporation was to have the power to maintain the hotel already in operation at Perry Springs, Pike County, Illinois, and also to build a railroad that would pass "as near to Perry Springs as may be practicable."

Indians called its "mighty medicine." Eventually Benjamin and his fellow investors in the Perry Springs Hotel and Railroad Stock Company, which included his father, reputedly spent $100,000 in building what they soon claimed was the finest natural mineral springs resort in America. Surrounding its enormous more-than-100-bed, four-story hotel was a collection of cottages and extensive gardens. Three main springs (lime, iron, and sulfur) ran to the baths to cure what its proprietor proclaimed was an almost endless list of diseases. Promotional brochures advertised chalybeate waters, that is, waters loaded with iron oxide, opening up to skeptics their alternative designation as rusty waters.[13,14]

In the beginning Ben remained much assisted by his father, WWWI, who lived to be 80 before he died in 1874, widely acknowledged as

[13] **W.W. Watson & Son advertised on January 23, 1865, that they had bought the Mineral Springs in Pike County and that the confectionary and ice business "will be speedily brought to a close." Offered for sale were the business, including 600 tons of ice, and several buildings and houses in Springfield, as well as 820 acres of land.**

[14] **An 1876 ad for the hotel offered "all rational amusements; dancing every evening, German every Wednesday evening." It also noted that W.W. Watson "will be found in the office."**

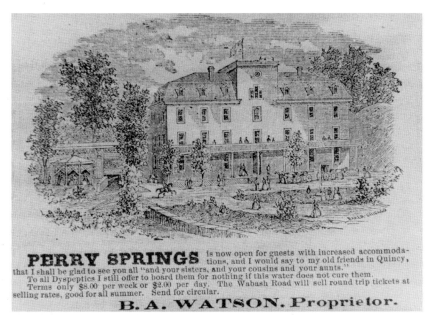

PERRY SPRINGS is now open for guests with increased accommodations, and I would say to my old friends in Quincy, that I shall be glad to see you all "and your sisters, and your cousins and your aunts." To all Dyspeptics I still offer to board them for nothing if this water does not cure them. Terms only $8.00 per week or $2.00 per day. The Wabash Road will sell round trip tickets at selling rates, good for all summer. Send for circular.
B. A. WATSON. Proprietor.

Advertisement for the Perry Springs Hotel that appeared in the Quincy Daily Herald *in 1865. The hotel's rate was $8 per week or $2 per day—and Ben promised guests: "To all dyspeptics I still offer to board them for nothing if this water does not cure them."*

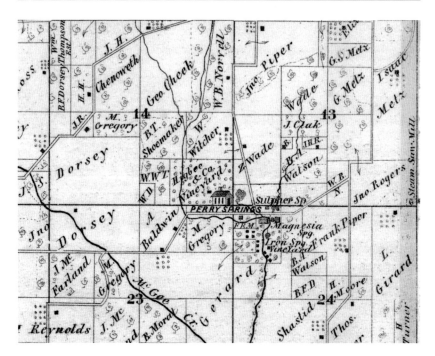

An 1878 receipt in Ben Watson's hand, including a $12 charge for mineral water.

Map of Perry Springs showing two plots of land owned by B.A. Watson.

This photo of Perry Springs was taken on August 12, 1894.

among Springfield's most prominent citizens. In his newspaper obituary, he was remembered as a man of sterling integrity whose many Christian virtues and estimable traits of character entitled him to the high place he held in public esteem. Long a member of the Baptist communion, he was in effect the founder and builder of the North Baptist Church. Two handsome, large memorials mark his and his wife's burial in the Oak Ridge Cemetery, where Lincoln a decade before was also interred.

For many years his son Benjamin's grand creation of Perry Springs flourished, being crowded each season with visitors from Illinois and neighboring states. There Ben maintained his proprietor's role until he sold out his last interests in 1890. For the rest of his life, which ended in 1901 at age 83, he lived in the Chicago region, where he helped his son WWWIII manage his real estate portfolio. Upon his death, his body was removed to Springfield, where a large monument in the Oak Ridge Cemetery placed next to that of his father marked out the burial sites for him and his wife and children.

In 1871, WWWIII (Welly), the first of Ben's eight children and my Dad's grandfather, made a highly successful marriage to Augusta Crafts Tolman, the talented pianist daughter of the St. Charles, Illinois, banker Thomas Frothingham Tolman and his wife Maria Paddock. Both of her parents were born in Craftsbury, Vermont, founded in 1781 by her

Funerals.

Watson—The remains of Benjamin A Watson, who died at his home in Chicago, will arrive here this morning. The funeral will occur this afternoon at 2:30 o'clock, from the residence of Dr. J. LaForest King, 613 South Fifth street. The burial will be private, and the remains interred at Oak Ridge cemetery.

The deceased was 83 years of age, and a former resident of Springfield. While here he was proprietor of a large confectionary store. He is survived by one son, W. W. Watson, of Chicago, and four daughters, Mrs. Emily Carey, of Pittsfield, Ill.; Mrs. Julia Smith, of Marshalltown, Ia., and Misses Hettie C. and Fannie Watson, of Chicago.

B.A. Watson's obituary notes that he was survived by his son W.W. Watson of Chicago and four daughters—Mrs. Emily Carey, Mrs. Julia Smith, Miss Hettie C. Watson, and Miss Fannie Watson.

William Weldon Watson III (1847–1913), grandfather of James D. Watson, Sr.

Burial monuments for the Watson family at Oak Ridge Cemetery in Springfield, Illinois. Four children survived W.W. Watson—B.A. Watson of Perry Springs, Mrs. T.S. Little, Mrs. Noah Divelbiss, and Mrs. John Ives. He also was survived by three sisters living in Morristown, New Jersey, as well as a large number of grandchildren and great-grandchildren.

grandfather, Ebenezer Crafts. For the next eight years, WWWIII, following in his father's footsteps, was the proprietor of a resort hotel on Fox Lake 30 miles northwest of Chicago, where his first sons—WWWIV (b. 1872) and my grandfather Thomas Tolman (b. 1876)—were born. Then WWWIII and his still-growing family moved north to the summer resort town of Lake Geneva, Wisconsin, where they lived in a large farmhouse just outside town. Lake Geneva first rose to real prominence after the great Chicago fire of 1871, when many prominent Chicago residents temporarily stayed there all year. There Augusta bore five more children, only three of whom survived childhood: Dudley Crafts (b. 1885), Bertram (b. 1887), and Garnett (b. 1891). In 1879 WWWIII leased Lake Geneva's grandest hotel, "The Whiting House," lengthened its season from May to November, and charged $2.50 for a night's stay. In this way WWWIII was for 15 years one of his town's most prominent citizens. His highly successful Lake Geneva hotelier career, however, came to a sudden end when the Whiting House burned to the ground from a fire in early July of 1894.[15]

Augusta Crafts Tolman (1852–1927) and WWWIII married on October 25, 1871. They had seven sons, five of whom survived, including James D. Watson Sr.'s father, Thomas Tolman Watson.

[15] A July 9, 1894, report of the fire describes the Whiting House as the largest hotel on the lake and the loss to the Whiting estate estimated at $25,000. The four-story wooden building was reduced to ashes within an hour, according to the report.

Again focusing on his real estate interests in the Chicago area was the obvious way to ensure that he and Augusta in their later years did not

B.

[7—296.]

Note A.—The Census Year begins June 1, 1879, and ends May 31, 1880.
Note B.—All persons will be included in the Enumeration who were living on the 1st day of June, 1880. No others will. Children BORN SINCE June 1, 1880, will be OMITTED. Members of Families who have DIED SINCE June 1, 1880, will be INCLUDED.
Note C.—Questions Nos. 13, 14, 22 and 23 are not to be asked in respect to persons under 10 years of age.

SCHEDULE 1.—Inhabitants in *Village of Geneva*, in the County of *Walworth*, State of *Wisconsin* enumerated by me on the *5th* day of June, 1880.

M. D. Cowdery, Enumerator.

	Name	Color/Sex/Age	Relationship	Civil Condition	Occupation	Health	Education	Nativity (self)	Nativity (father)	Nativity (mother)	
20	122 131 Knappler William	W M 40			Boat Builder			England	Eng	Eng	20
21	Mary A	W F 40	Wife		Keeping House			England	Eng	Eng	21
22	Edward H	W M 14	Son		At School			Eng	Eng	England	22
23	Albert E	W M 13	Son		At School			Eng	Eng	Ireland	23
24	Annie F	W F 7	Daughter					Wis	Eng	Ireland	24
25	Ada E	W F 5	Daughter					Wis	Eng	Ireland	25
26	123 132 Duland James	W M 27			Servant at Hotel			Missouri	Missouri	Missouri	26
27	Dolly	W F 23			Keeping House			Missouri	Missouri	Missouri	27
28											28
29	Watson William	W M 33			Hotel Proprietor			Illinois	Tenn	Ill	29
30	Augusta C	W F 28	Wife		Keeps the House			Maine	Vt	Vt	30
31	Willie	W M 7	Son		At School			Illinois	Ill	Maine	31
32	Thomas S	W M 4	Son					Illinois	Ill	Maine	32
33	Tolman Thomas H	W M 60	Clerk		Hotel Clerk	Inflammation in tract		Vt	Vt	Vt	33
34	Tolman Augusta	W F 54			Housekeeper			Vt	Vt	Vt	34
35	Arnold Charles C	W M 30	Telegraph Op'r		Telegraph Operator			Wis	Pa	N.Y.	35
36	Butler William	B M 47	Cook		Hotel Cook			Penn	Md	Penn	36
37	Rogers Thomas	W M 56	Servant		Hotel Storekeeper			Ireland	Ireland	Ireland	37
38	Oats Anne	B F 23	Servant		2nd Cook			Alabama	Ala	Ala	38
39	Noxsen Amanda	B F 18	Servant		Servant			Alabama	Ala	Ala	39
40	McNamara Annie	W F 37	Servant		Servant			Ireland	Ireland	Ireland	40
41	Leslie James	W M 26	Servant								41
42	Pickley Jerry A	W M 31	Servant		Bar Keeper			Md	Md	Ireland	42
43	Lewis James	W M 19	Servant		Bar Keeper			Vt	Vt	Vt	43
44	Gaffney Emma	W F 22	Servant		Servant			Wis	N.Y.	N.Y.	44
45	Casey Ella	W F 14	Servant		Servant			Wis	Ireland	Ireland	45
46	Casey Josey	W F 24	Servant		Servant			Wis	Ireland	Ireland	46
47	Hunter Sarah	W F 52	Boarder		Boarding			Ireland	Ireland	Scotland	47
48	Hunter Blanch	W F 20	Boarder		Boarding			Canada	England	Ireland	48
49	Scheurman Theodore	W M 49	Clerk		Hotel Clerk			England	Germany	England	49
50	O'Brien Annie	W F 19	Waiter		Seamstress			Ill	Ireland	Ireland	50

NOTE D.—In making entries in columns 9, 10, 11, 12, 16 to 23, an affirmative mark only will be used—thus ✓, except in the case of divorced persons, column 11, when the letter "D" is to be used.
NOTE E.—Question No. 12 will only be asked in cases where an affirmative answer has been given either to question 10 or to question 11.
NOTE F.—Question No. 14 will only be asked in cases when a gainful occupation has been reported in column 13.
NOTE G.—In column 7 an abbreviation in the name of the month may be used, as Jan., Apr., Dec.

1880 United States Census indicating the extent of WWWIII and Augusta's household in Geneva, Wisconsin, site of the Whiting House hotel.

Whiting House hotel. The 60-room hotel was built in 1873 at a cost of $30,000 and was located on the boat lagoon of Lake Geneva. A contemporary account describes the hotel as "elegantly appointed, its cuisine perfect, and every convenience supplied" and "a popular summer rendezvous of merry-makers." W.W. Watson is described as "the urbane proprietor" with the "requisite faculty of rendering his guests content."

drift toward penury. WWWIII, however, knew that the grandeur of his Whiting House days would be lost forever if he did not somehow go for gold. Somehow he had to repeat his father Ben's accomplishments during the 1848–1849 California Gold Rush. Learning of a prospective gold find in Honduras in the mid-1880s, he made a significant speculation in below-ground gold mines (as opposed to panning for gold nuggets in the California Gold Rush). In doing so, he in effect became a mining engineer with knowledge on his side. His son Dudley Crafts long afterward would tell outsiders that his father was by then more a mining engineer than a hotel keeper.[16,17]

Then regretting that none of his five sons would likely have the technical education needed to assist him as real partners for his mining hopes, a decade later in 1900, Welly covered the costs of Dudley's first-

[16] The discovery of gold in Honduras was reported in Chicago as early as January 1886. Chicago businessmen were quick to organize, including W.W. Watson and B.A. Watson. A state incorporation license was issued in 1888 for the Guyape Gold Mining Company of Honduras with capital stock of $1 million. The incorporators were J.H. Secrist, R.S. Price, B.A. Watson, and W.W. Watson. Another license of incorporation for La Victoria Mining and Milling company in 1887 at Chicago with capital stock of $1,500,000 listed William W. Watson as one of its incorporators.

[17] W.W. Watson and B.A. Watson attended a reception for the Vice President of Honduras in August of 1887 in Chicago. A report in the *Daily Inter Ocean* of Chicago adds that, "W.W. Watson accompanied by Mrs. Watson on piano, whistled a few very difficult airs."

Ad for WWWIII's Restaurant Francais (complete with misspelling) in Chicago.

WWWIII's endorsement of Paine's Celery Compound appearing in the Daily Illinois State Register.

ALL CHICAGO KNOWS HIM.

W. W. Watson, Leading Real Estate Man, Restored to Health by Paine's Celery Compound.

Chicago, March 24.—Mr. W. W. Watson's reputation throughout the west for unerring judgment in the valuation of land has made him foremost among the most conservative, careful class of investors in Chicago.

Unlike many hard driven business men, the owner of "Alpine Heights," that splendid suburb of Chicago, has not neglected his health on account of his exacting business. The following unrequested statement from Mr. Watson shows how consistent with his life-long, careful, conscientious and successful business habits has been his attention to getting well. He states in the Times-Herald:

"Upon the recommendation of a friend, I used Paine's Celery Compound for headaches, constipation, indigestion and loss of sleep, and found it all it was recommended to be. I suffer no more from headaches, sleep soundly at night and am now in perfect health. This is the only medicine that I have ever taken for these complaints, which has benefited me at all.

W. W. WATSON.
225 Dearborn street.

has led to worrying, fretting and despondency, that need only Paine's Celery Compound to banish the unhealthy atmosphere and make things bright and cheerful again.

It is the only spring remedy universally prescribed by physicians. It makes people well by giving them a hearty appetite and a relish for their food. Hard-worked men and women, the nervous, weak and debilitated, get new strength, fresh nervous energy and a purer, more vigorous blood supply from Paine's Celery Compound.

This most valuable nerve and brain invigorator and restorer practically demonstrates the life-long conviction of its eminent discoverer, Professor Phelps, M. D., LL. D., of Dartmouth college. Professor Phelps was for a long time convinced that sound nutrition was the keystone of firm health, and that where there were signs of poorly nourished nerve tissues, and of thin, pale-colored blood, some means must be devised to supply these deficiencies briskly and rapidly. Professor Phelps prepared Paine's Celery Compound on this basis. It has proved an invigorator, strengthener

WWWIII and Augusta Crafts Watson with younger sons Dudley Crafts (lower right*),*
Garnett (left*), and Bertram* (seated middle*).*

year tuition at the nearby Illinois Institute of Technology (IIT). The Institute was located just over a mile from WWWIII and Augusta's home,
now back in Chicago at 4156 Berkeley Avenue, slightly north of the
Kenwood area where many of Chicago's most successful German-Jewish merchants were building large, wooden, Victorian-style homes.

Dudley Crafts, although highly intelligent, was never destined, however, to be a chemical engineer. His IIT advisors soon discovered that he
was likely to be much more successful as an artist (he had an uncanny
ability to catch the likeness of anyone he chose to sketch). At the same
time, he was being taught the piano by his mother, who wanted him to be
a professional player carrying on her own glory. Initially he chose neither, becoming an actor in a professional touring company that took him
in 1904 at age 19 to St. Louis, where the World's Fair was in progress,

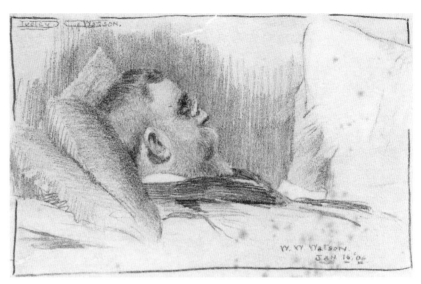

Dudley's 1906 sketch of his father shows his artistic skill.

A young Dudley Crafts Watson.

giving him his first glimpse of the French Impressionist painters. They so fascinated him that he made up his mind that he did not want to act but was born to paint. With $100 borrowed from his father for tuition, he entered the Art School of Chicago's Art Institute, working as a guard at night and posing for sketching classes to help pay his way. After only two years as a student, he was made an instructor and remained in that capacity for three years, during which he married Laura Hale in 1908.

Unfortunately, his father Welly's efforts to make a late-life fortune from gold mines in Honduras never returned even a penny to him and his fellow investors. The jig was up even before the Whiting House burned to the ground in 1894. His final 12 years of life (1901–1913) saw him join the *Chicago Daily News*'s free lecture staff, where his subjects included

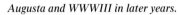

Augusta and WWWIII in later years.

This ca. 1905 photo of WWWIII was taken by a Chicago Daily News *photographer.*

"Honduras: Gold from Mine to Mind" and "Ranch Life" about living outside Lake Geneva during his ownership of the Whiting House Hotel. For his burial, his body returned to the Lake Geneva community that witnessed his most rewarding years.

CHAPTER 2

Tolman's Fateful Plunge

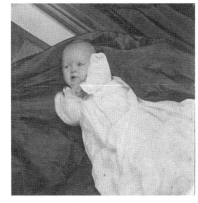

The infant James Dewey Watson.

My father, James Dewey Watson, known to all his friends as Jim Watson, was born on December 13, 1897, in the iron ore town of Eveleth, Minnesota, near the center of the Mesabi Range, some 60 miles north and east of Minnesota's Lake Superior port city of Duluth. A year earlier Thomas Tolman, his father, had moved to the Range soon after his around 1895 marriage at age 20 in Lake Geneva to the similarly aged Nellie Dewey Ford. How Tolman might best carve out a future business life that would support his growing family then was not obvious. Joining his father, William Weldon Watson III (WWWIII), in the resort hotel business stopped being in the cards once he was no longer the proprietor of the Whiting House. Common sense dictated that his father devote virtually all of his non–gold mine efforts to exploiting his seemingly instinctive capacity to spot good real estate deals in the Chicago area. At best, only minor sums might become available over the next several years to help his children move into their own businesses. Already Tolman's older brother William Weldon Watson IV (WWWIV, or Willy; 1872–1958) sought a non-hotel career through moving to the Mesabi Range. There hard work and determined common sense could rapidly make newcomers financially comfortable. Unfortunately, Tolman's excitable up-and-down charming personality was not what the Range needed. Bookkeepers, no matter how charming, are not hired to take chances.

Fortunately, the early death of Nellie's lumberyard-wealthy, widower father, Delavan Ford (1825 or 1830/1832–1905), gave Tolman and Nellie the financial means to move back to the Midwestern social milieu from which they had come, after five grueling years on the Range. Wisely they

James D. Watson, Sr.'s father, Thomas Tolman Watson (1876–1930).

Nellie Dewey Watson (1875–1936), James D. Watson, Sr.'s mother, in a photo taken ca. 1900. The infant in the photo could be either James D. Watson, Sr.'s brother Tom (b. 1901) or William (b. 1899).

Nellie's father, Delavan Ford (c. 1830–1905), with his grandsons Thomas Tolman Watson II, James Dewey Watson, and William Weldon Watson.

chose to relocate to the already prosperous Protestant Chicago suburb of La Grange, 16 miles southwest of Chicago's Loop. There Nellie bought a large white colonial revival-style house next to its golf course. By then she already feared that Tolman as a stockbroker would someday make a speculative bet on stocks that he had not the capital to cover if they lost too much of their value. So Nellie never let her husband have any control over her new inheritance, their large home being solely hers.

Nellie was by far the more intellectual of Dad's parents. She personally bought most of their La Grange home's books, some dating to before her marriage, including a volume of Keats' poems and an 1892 edition of Sir Thomas More's long-celebrated *Utopia*, in which he described an imaginary island enjoying perfection in law and politics. Dad's father, Tolman, although less bookish than his wife, owned a highly detailed *Short History of the English People* that he had used as a student in the Lake Geneva school system. In 1909 Nellie gave Tolman a copy of the much-celebrated 1880 novel *Ben-Hur* by Lew Wallace. As a devout Episcopalian, Nellie later put money in her will toward

Stockbroker Thomas Tolman and Nellie's LaGrange, Illinois, house "Whitehouse" in 1915. When their financial fortunes soured, they sold the house and moved to a large apartment near the University of Chicago campus.

Books from Thomas Tolman and Nellie's library.

Young James D. Watson, Sr., with baseball gear.

the purchase of highly prized, stained glass windows in her hometown church. Although Tolman was from a long Baptist heritage, he sensibly adopted the Episcopalian religion of his wife. She, however, never was comfortable with the big-money Republican Party and always voted for the Democratic candidate for the Presidency.

A major attracting feature of La Grange was its high-quality Lyons Township High School, which Dad attended from the fall of 1911 to the spring of 1915. There, through the encouragement of its Biology teacher, he and his classmates Ed Hulsberg and Sidney Wade became ardent young ornithologists, regularly making several-mile-long bird walks after school and later writing up their observations in essay forms likely also seen by their English teachers. Already then Dad enjoyed

List of Warbler sightings compiled by James D. Watson, Sr., in the spring of 1914, while he was still a high school student.

JAMES D. WATSON
WARBLERS SEEN IN 1914 (SPRING)

| Month / Date | BLACK and WHITE | PROTHONOTARY | BLUE WINGED | GOLDEN WINGED | NASHVILLE | TENNESSEE | PARULA | CAPE MAY | YELLOW | BL. TH. BLUE | MYRTLE | MAGNOLIA | CERULEAN | CHESTNUT SIDED | BAY BREASTED | BLACKBURNIAN | BLACKPOLL | BL. TH. GREEN | PINE | PALM | OVEN-BIRD | WATER THRUSH | CONNECTICUT | MOURNING | MARYLAND YELLOW THR. | WILSON | CANADIAN | RED START |

Feb. 24, 1914
31- S.W.- W.
Snow

A Trip to the North Woods

The snow started to melt this afternoon for the first time this month. There was still about eight inches of snow left. The birds are having a hard time getting food as they can not reach the ground. The snow has not left the ground this month. Ed and I started out about 2:15 and got back at 5:00. We made the usall trip per the woods. The creek was frozen solid so we had a good highway to walk upon which made it easy.

When we were going down Ogden Ave. we were surprised to hear some birds singing to the right of us in the field. We thought they were Horned Larks but were not sure. We got right on top of them before we could see the birds. They were Horned Larks. This is the first time we have heared them sing this year. It was a short warble utered constantly the whole time we were there. We then went on and heared two Downys pecking away on the limbs of the trees along the road. One of them was evidently in the hole because we could not find him. We did not see the Red Head.

We then went on to the road leading to the creek. Near the haystack we ran into a flock of Tree Sparrows. They were flying from the haystack to the brushpile utering their "chip" continually. We noticed the absence of the Juncos which were with us last year at this time and place. We heared the rolling note of the woodpecker in the distance which sounded like the note of a Tree Frog which I heared many nights in summer.

Going down to the creek we started east back per the woods. When we were walking along we found a good many tracks of birds who had probably come up under the snowbank at night to escape the wind and storm. As we were walking around the bend in the creek we heared the note of a sparrow in the distance in the dark woods. When we found him we thought at first that he was a Redpoll but we soon saw the black spot that the Tree Sparrow has in the center of his breast. He had a remarkable bright red head.

Around the next bend we saw two Chicadees. They were back of us and we determined to find them. We soon did and got very close to them. Ed can whistle the note of the Chicadee so he will answer him every time. His "chic·a-dee-dee"sounded great in the cold woods. Off to the right we heared two woodpeckers making an awful noise. We thought at first that they were flickers on-account-of their note which sounded exactly like that of the Flicker. We soon found out that they were Hairy's. They were chasing each other per the oak leaves.

We then plowed per the snow to the academy fence where we heared another woodpecker tapping in the distance on a hard tree. Here Ed climbed over a fence to get a cocoon but found out that it was empty. He has got some beauties in the last few weeks. We heared a good many crows in the woods and later two. We also saw a Downy in the distance. As we were starting back home we heared a Nuthatch call a few hundred feet away. The Screech Owl was in his hole.

March 11, 1914
Warm—N—38
North Woods

Ed, Bill and I started out about 2:20 and got back 5:39. We went way down to Fulersburg along the creek and then north to the Proviso Road at Twenty-second St. It was very good walking although slightly muddy. We haven't made this trip since before Christmas it being a rather long one.

Cutting across fields we saw plenty of Crows and three Horned Larks which were feeding in a plowed field. They were just the color of the dirt and we had a hard time finding them. They were singing very beautifully tho not very loud. They are the first birds to sing in the spring, starting usually the last week in February. When we got up to the road again Ed heared the Quail whistle off in the woods. He [answered] but we did not hear him again tho we cut through the patches.

On the road leading to the creek we ran into a very large flock of Tree Sparrows. They were all warbling which sounded very beautiful. One of them gave us his full song. They seem to be gathering in flocks now. A Downy was in the tree above us getting the grubs and eggs. The absence of the Juncos reminded me of the same period last year when we saw so many. We went down to the creek and then started north. Across the Proviso Road there was another flock of Tree Sparrows. Their "chip" sounded very loud and clear. When we got down to the swamp near the bend of the creek we were surprised to see two Brown Creepers winding their way up the trunks of an Oak Tree. Their little wing [?] note which they gave very often sounded very pritty.

From hear we continued north west to Twenty-second St. Hear is where we saw the Flicker near Christmas time. There was not much of anything here except Crows of which I counted about a dozen. They would fly very swiftly around over the tops of the trees, and then glide gracefully down to the top of one of them, all the time giving their "caw, caw, caw." They also seem to be flocking.

We then passed the large swamp which ought to be fine for water birds this spring if it fills with water. On the corner of 22nd St. and Proviso Road there are a great many Pine Trees. In one of them we saw a Screech Owl sleeping. He was very plain being the brown phase of the Screech Owl. He did not seem to notice that we were there at all. From here we cut through the woods to the creek. Another large flock of Crows passed over (15). They were going towards the creek. We heared the Hairy Woodpeckers rattle in the distance among the Oak Trees. The only bird we saw near the creek was a solitary Herring Gull which sailed majestically over us. We did not see the Chickadees or the Cardinals. When we started out we thought we might see some of the early migrants but we did not. The Meadow Larks and Robins are due in a few days.

March 16, 1914
West of School
La Grange

 Very warm again this morning with the sun shining brightly. Got up early and went west of school before school. The Robins were very plentiful this morning a good many singing. Saw my first Bluebird. He was across from Bassfords and appeared very nervous tho he was warbling his beautifull song. He seems to carry the sky on his back, the blue seems to be just the color of the sky of this morning. A few English Sparrows came along and started to make life miserable for the poor Bluebird. These birds are never very numerous around La Grange tho there are allways quite a few out in the woods.

 Cutting across the fields I heared the Redwings and soon saw them in the swamp with their beautiful scarlet shoulder patches. I heared their "Oak-a-lee" way over on Brainard Ave. They soon flew on toward Western Springs. It seems hardly yesterday when I last saw those birds in the fall. We used to see large flocks of them then. Also saw about ten Juncos in the Willow Patch. They were very beautiful, their slate breast contrasting nicely with the white belly. They are becoming more common every day. The insects have been flying around since last Friday the 13th. Saw a few moths and bees. Ed said he heared the frogs croaking yesterday.

March 29, 1914
65—S.W.—Wet
West To Ogden

 Today was another warm day the [temperature] being about sixty-five and the humidity very high. Started to rain this afternoon just as we got out to Ogden Ave. As it was I saw twenty-four [different] kinds without going to the North Woods. It has rained pretty hard for the last four or five days and the swamps and fields are being filled up rapidly. The Hermit Thrushes arrived today there being three of them over Ludwigs [?]. I saw them this morning for the first time. I thought for the instant that they might be Fox Sparrows but their actions and red tails so gave there [*sic*] identity away. They act alltogether diferent from the Fox Sparrows. They are very much quieter and hold their head high in the air in a lordly sort of away. They were very silent. Last year I did not see them untill the 2nd of April. They will probally become common in a day or two. This morning west of school I heared the beautiful song of the Purple Finch. It was a loud sweet ear __[?]. This is the first time I have ever heard the song of the Purple Finch. The Juncos were very abundant and all singing in chorus. At this time I saw a Sparrow Hawk fly swiftly over towards the woods. There was a large flock of Cowbirds in the swamp. This afternoon I started out to the Woods with Ed butt it started to rain and we came back. We saw two other new birds here the Pheasant and Savannah Sparrow. We saw the Pheasant here last fall in the same place (See Card Index). On this road we saw our first ant hill of the season. There were a great many worms crawling around today.

Brothers James, William, and Thomas Watson on the 1912 trip to Estes Park in the Colorado Rocky Mountains.

expressing himself through well-chosen words and phrases. That summer Dad's lifetime list of birds went up dramatically when he and his brothers Bill, Tom, and Stanley went by train with their parents to Denver on their way to Estes Park in the high Rockies. There they stayed at a dude ranch and learned to ride horses.

Sundays then, however, were not for birds, with Dad receiving prizes for best Episcopal church attendance for six successive years. In the fall of 1915 he left home to enroll at Oberlin, the high-principled, academically much admired, liberal arts college located 35 miles southwest of Cleveland. A main reason for his going there was the presence of its Biology teacher, the then-noted ornithologist Lynds Jones, whose 1895 course in Ornithology (the scientific study of birds) was the first such college-level offering in the United States.[1,2] Even before Dad was born, Jones had established and become the first editor of the birding journal *The Wilson Bulletin*. The weekend-long excursions that Jones led to the migratory, bird-rich Put-in-Bay Island just off Lake Erie's shores,

[1] One of Jones's former students described him as "a teacher who places the world of nature before you, and with a few guiding remarks expects you to make the discoveries for yourself."

74   THE WILSON BULLETIN—No. 95

THE MAY BIRD CENSUS.

Particular interest attaches to the 1916 May census because, throughout the central states, the weather conditions favored the halting of many of the smaller birds as well as many of the larger ones during the last week in April and the first half of May. In the vicinity of Oberlin vegetation was backward, the weather prevailingly cold, with occasional warm days, and the precipitation, while not excessive, was spread over many days. The appended list for the first of May shows that the warblers as well as other small birds, were ahead of the season as well as being ahead of schedule. The list for May 8 and 15 shows that there had been little change in the bird population except the arrival of more species—only eight species having left, the rest of those not seen on the 8th as well as the 15th being breeding birds.

Oberlin, Ohio, May 1st, 1916. Chilly morning, warming to about 80 by noon. Cloudy early, clearing until about 1:00 p. m., then clouding again to showers and a steady rain by mid-afternoon. Wind strong S. W. About Oberlin early morning, Berlin Heights to Lake Erie via Old-woman's Creek 8:00 to 10:15 a. m., Rye Beach to Cedar Point 10:30 to 3:30. Lynds Jones, Max. de Laubenfels, and James Watson, until 10:15 a. m., Jones leaving the party then.

Pied-billed Grebe, 1; Horned Grebe, 1; Herring Gull, C; Bonaparte Gull, 2; Common Tern, 10; Black Duck, 10; Lesser Scaup Duck, 4; Solitary Sandpiper, 1; Woodcock, 1; Spotted Sandpiper, 25; Killdeer, 10; Coot, 25; Florida Gallinule, 3; Sora, 5; Virginia Rail, 1; Bittern, 3; Great Blue Heron, 1; Green Heron, 2; Bobwhite, 3; Mourning Dove, 10; Turkey Vulture, 3; Marsh Hawk, 10; Sharp-shinned Hawk, 15; Red-tailed Hawk, 2; Red-shouldered Hawk, 5; Broad-winged Hawk, 2; Bald Eagle, 2; Sparrow Hawk, 10; Barred Owl, 2; Screech Owl, 1; Belted Kingfisher, 3; Hairy Woodpecker, 2; Downy Woodpecker, 5; Yellow-bellied Sapsucker, 2; Red-headed Woodpecker, C; Red-bellied Woodpecker, 4; Northern Flicker, C; Whip-poor-will, 3; Chimney Swift, C; Kingbird, 3; Crested Flycatcher, 2; Phœbe, 1; Least Flycatcher, 3; Blue Jay, C; Crow, C; Bobolink, 5; Cowbird, C; Red-winged Blackbird, C; Meadowlark, C; Baltimore Oriole, 4; Rusty Blackbird, 2; Bronzed Grackle, C; Goldfinch, 3; Vesper Sparrow, C; Grasshopper Sparrow, 3; White-throated Sparrow, C; Tree Sparrow, 1; Chipping Sparrow, C; Field Sparrow, C; Slate-colored Junco, 1; Song Sparrow, C; Lincoln's Sparrow, 2; Swamp Sparrow, 3; Towhee, C; Cardinal, 10; Rose-breasted Grosbeak, 3; Indigo Bunting, 3; Scarlet Tanager, 1; Pur-

James D. Watson, Sr.'s Oberlin College Professor of Ornithology and Zoology Lynds Jones (right), shown here on an August 1900 trip with William L. Dawson (left) to Washington State.

The May 1916 Oberlin bird census published in The Wilson Bulletin *of June 1916 features sightings by Lynds Jones, James Watson, and Max de Laugenfels on May 1 and additional sightings by James Watson and five others on May 15. (Figure continued on following page.)*

The May 1916 Oberlin bird census continues from previous page.

ple Martin, C; Cliff Swallow, 3; Barn Swallow, C; Tree Swallow, 1; Bank Swallow, C; Rough-winged Swallow, 10; Cedar Waxwing, 12; Migrant Shrike, 1; Warbling Vireo, 4; Blue-headed Vireo, 5; Yellow-throated Vireo, 1; Black and White Warbler, 5; Blue-winged Warbler, 3; Golden-winged Warbler, 1; Worm-eating Warbler, 1; Nashville Warbler, 10; Orange-crowned Warbler, 1; Yellow Warbler, C; Black-throated Blue Warbler, 5; Myrtle Warbler, 10; Magnolia Warbler, 1; Chestnut-sided Warbler, 3; Blackburnian Warbler, 3; Black-throated Green Warbler, 10; Palm Warbler, 15; Oven-bird, 5; Water-Thrush, 1; Louisiana Water-Thrush, 1; Maryland Yellow-throat, 3; Redstart, 1; Pipit, 5; Catbird, 5; Brown Thrasher, 5; House Wren, 5; Winter Wren, 2; Long-billed Marsh Wren, 2; White-breasted Nuthatch, 5; Red-breasted Nuthatch, 1; Tufted Titmouse, 10; Chickadee, 3; Golden-crowned Kinglet, 3; Ruby-crowned Kinglet, C; Blue-gray Gnatcatcher, 5; Wood Thrush, 10; Veery, 2; Olive-backed Thrush, 5; Hermit Thrush, 15; Robin, C; Bluebird, C. Total species, 117.

At Oberlin, on May 8, another study was made by H. W. Baker, Lynds Jones, Max de Laubenfels, and Lester Strong, covering much the same ground as on the first, but extending the studies to the east end of the Marblehead peninsula across the bay from Sandusky. The time spent was from 3:30 a. m. to darkness. The day was fair but with a brisk south-west wind.

Another all day study was made on May 15 in the same general region, by Mr. George L. Fordyce and John Young of Youngstown, Lynds Jones and Max de Laugenfels, and Harry G. Morse and James Watson, who worked only in the vicinity of Oberlin. The first four named spent the early morning at Vermilion, Mr. Morse the early morning at Huron, and the five then worked together at Rye Beach for a half hour, then crossed to the Marblehead peninsula and worked there until 3 p. m. There was a thunder-shower in the early morning, then a clearing and warm day, with little wind.

It seems best to arrange the species in tabular form, including those found at Youngstown on May 12, when an all day study was made by Messrs. Fordyce, Jones, de Laubenfels, Leedy, Murie, Rogers, Todd, Warner, and Young. In this study the parks, woods and artificial lakes in the region of Youngstown were visited. The day was fair, with little wind.

	Oberlin May 8	Oberlin May 15	Youngstown May 12
Pied-billed Grebe	1		1
Loon			3
Herring Gull	c	c	
Ring-billed Gull	3	3	

[2] Jones, whose main scientific interest was bird migration, was an early proponent of bird counts.

45 miles away, were the highlights of Dad's Oberlin ornithology experience.

As befitting a Protestant Evangelical college, Oberlin made Bible studies an obligatory first-year experience. Throughout his life Dad kept an earlier paper that he had composed for it, likely signifying that its writing gave him pleasure. His iconoclastic, free-thinking Bible course was taught by the Oberlin- and Yale-educated theologian, William (Will) James Hutchins (1871–1958), himself the son of an equally academic theologian, Robert Grosvenor Hutchins (1838–1921), whose 50-year career concluded at the long racially integrated Berea College of Kentucky. Will had moved back to Oberlin as the Holbrook Professor of

William James Hutchins (1871–1958) on his inauguration as the fourth President of Berea College on October 22, 1920. His son Francis S. Hutchins succeeded him in this post in 1939.

Bible I   James D. Watson
                                                   Sept. 25, 1915

The First Epistle General of John

I.   I have read the First Epistle of John.

II.   I think that John I, II, III have all been written by the same author. A good many of the words in the three Epistles are identical or very similar. For both the First and Second Epistle, the author calls all people who do not confess that Jesus Christ is come in the flesh, deceivers and antichrist. (First Epistle John—Chap. 2 Ver. 22) (Second Epistle John—Ver. 8). In both these Epistles the author says "that all those who do acknowledge the Son hath the Father also, and all those who denieth the Son, the same hath not the Father." (First Epistle John—Chap. 2-3 Ver. 23, 24) (Second Epistle John—Ver. 9.) Also the general theme of thought seems to be very similar in all three Epistles. The beginnings and endings of the Epistles are very much alike. In all three Epistles he tells his readers to love one another, and believe in the Doctrine of Jesus Christ. I do not see any reason for suspecting different authorship.

III. The author of the First Epistle of John has a great many purposes in writing it. He brings these out quite clearly in the text. For the first chapter he brings out the point that we should confess our sins and that God will cleanse us from all unrighteousness. For the second chapter he urges them to keep the commandments and to love their brothers. He says that those who hate their brothers are in darkness. The strongest point that he brings out in this chapter is "To love not the world." He says — "If any man love the world, the love of the Father is not in him." In the third chapter these same points are brought out again. In the fourth chapter he warneth them not to believe in all teachers, who boast of spirit, but to try them by the rules of the catholic faith. The last part of this chapter is also devoted to brotherly love. In the last chapter he says that Jesus is the son of God, able to save us and to hear our prayers, which make for ourselves and for others.

IV. In several places the author gives his conception of God. In Chap. 3 Verse 5 he says — "And ye know that he manifested to keep away our sin." God is therefore a sinless man. In Chap. 3 Verse 20 he says — "For if our heart condemns us, God is greater than our heart, and knoweth all things." God knows all things. In Chap. 4 Ver. 1 — "God is love, and he that dwelleth in love dwelleth in God, and God in him." In the last verse of the Epistle he says: — "This is the true God, and eternal life."

V. The author's conception of Jesus' relation to God and to men are: (1) Chapter 1 Verse 7: — ". . . and the blood of Jesus Christ his Son cleanseth us from all sin." (2) Chap. 4 Verse 15: — "Whosoever shall confess that Jesus is the son of God, God dwelleth in him, and he in God." This shows the relation of Jesus to God. (3) Chap. 5 Verse 12. "He that hath the Son hath life; and he that hath not the Son of God hath not life."

VI. "The World" as the author uses the term is used to express everything that has to do with evil—the lust of the flesh; the lust of the eyes; and the pride of life. The author says that "The world passeth away, and the lust thereof: but he that doeth the will of God abideth for ever." He writes if any man love the world, the love of the Father is not in him. They use the word nowadays in the same way.

VII. A great many things characterize a Christian according to the author. He must walk in the light and have fellowship with another. He must confess his sins and tell the truth. He must keep his word — "in him verily is the love of God perfected." He must love his brother and love not the world. He must not deny Jesus is the Christ. The rights and privileges of a Christian are: (1) Fellowship (2) Eternal Life

VIII. The author tells us that an antichrist is "He is antichrist, that denieth the Father and the Son."

IX. Some of the characteristic words of the Epistle are: love, light, life, truth, eternal life, fellowship, sin, cleanse, commandments, Father, lust, abide, brother.

Homiletics (moral preaching) in 1907 from Brooklyn, where he was the highly respected minister of the Bedford Presbyterian Church. He did not believe in original sin, implied guilt, or the absolute historicity of Genesis or Exodus.[3] When Will Hutchins left Oberlin to succeed his father as President of Berea in October of 1920, he worked tirelessly to further racial understanding.

[3] **Oberlin emphasized traditions of service and of educating the intellect and the character.**

Whether Dad at Oberlin had already begun to have doubts about the existence of God I don't remember his ever telling me. By then he was much under the influence of his teacher Will's son and a fellow freshman student, the unbelievably charismatic Robert Maynard Hutchins, who later came much into my family's life as the youngest-ever President of the University of Chicago. Possibly because as Will Hutchins' son he was expected to be well behaved, Robert Maynard Hutchins took joy in behaving seemingly out of control, stretching his community's tolerance for practical jokes to the limit, for example, by climbing the school powerhouse smokestack to paint his class's 1919 numeral. Campus rules early on became his *bête noir* as revealed by his widely circulated "Early Treatise on Education":

In Oberlin, I must not smoke.
I don't.
Nor listen to a naughty joke.
I don't.
To flirt, to dance, is very wrong.
Wild youths choose women, wine, and song.
I don't.
In Oberlin I must not wink at pretty girls
Or even think about intoxicating drink.
I don't.
I kiss no girls, not even one.
Why, I don't know how it is done.
You wouldn't think I'd have much fun.
I don't.

William's son Robert Maynard Hutchins (1899–1977) during his Oberlin years. An Oberlin classmate of James D. Watson, Sr., he later also figured in the educational life of James D. Watson, Jr.

The formidable Augusta Crafts Tolman Watson (1852–1927).

Laura Hale (1885–1962) and Dudley Crafts Watson (1885–1972) married in 1908.

Always saving Hutchins from career-destroying penalties were his high academic achievements, him seemingly being the best at any task he put his heart to. Dad was not so protected after being caught with Hutchins going off campus to smoke. Academically floundering after a prolonged scarlet fever–induced stay in the college infirmary, Dad was expelled from Oberlin at the end of his freshman year. His parents' marriage by then was marked by frequent quarrels over money. Tolman later was characterized by his artist brother Dudley Crafts Watson as a "plunger" at the crest of a wave or in the depths. Nellie, in contrast, was thought by Dudley to be a balanced wheel and mediator who tried to keep a tight rein on Tolman and devoted her life in every way to her sons. After a visit by Dudley and Tolman's mother, Augusta, to their La Grange home, she was reportedly in tears when she returned to her Chicago home, saying "Oh, I hope those dear boys can go away to school. The way their mother and father quarrel will ruin their lives."

Dudley's career, in contrast, was very much on course. By then he knew that his career would not be that of a painter but likely that of a very successful lecturer on art. Following his 1908 marriage to Laura Hale when they were both 23, he resigned from his teaching position at the Art Institute of Chicago to move with her to Lake Geneva, where he could paint full-time with monies hopefully coming from their sales giving them a steady income. Within six months, however, Dudley realized there were no streams of art lovers wanting to buy his landscapes or portraits. Fortunately his multifaceted talents as an art lecturer were soon also apparent to the citizens of the growing brewery-dominated city of Milwaukee, where plans already were being drawn up for its own Institute of Art. Within a year Dudley had become its first Director, a position he would continue to hold for the next 13 years. Soon he began leading groups of students to Europe to learn more about the greatness, past and present, of the past five centuries of European art. He had a group of 31 art students in Paris when World War I broke out. Detained for several weeks under mili-

An early Art Institute of Chicago art class. Dudley is at the back on the right-hand side.

tary orders, he hired a model for the group to paint mornings and arranged for French and ballet lessons in the afternoon.

Often Dudley would liven up his lectures with music, much of it performed by his highly talented second cousin, Beatrice Ives Welles, the granddaughter of WWWI's first daughter, Abigail. By then married in 1903 to the prosperous inventor Richard Hodgdon Head Welles, Beatrice lived in his hometown of Kenosha, Wisconsin, the first city of any size south of Milwaukee on Lake Michigan. Although both were of high intelligence and from prosperous backgrounds, their personalities and basic aims for life were very different. Beatrice abhorred alcohol, being a deeply committed Prohibitionist. In complete contrast, her husband Richard became increasingly addicted, becoming by 1912 hopelessly alcoholic. Unfortunately, their first child, a son also called Richard, was mentally ill and never achieved an independent life. Their second son, Orson, born

Orson Welles's mother, Beatrice Ives Welles (1882–1924). An accomplished pianist, she provided musical accompaniment for Dudley's art lectures.

a decade later in 1916, was highly unusual in a much better sense—so highly vocally precocious that he was declared a genius by 18 months of age. Although his mother wanted him to follow in her footsteps as a pianist, Orson's piano lessons when five and six years old did not portend extreme talent like that possessed by his mother, who once played with the Chicago Symphony Orchestra.

Unfortunately, my grandmother Nellie Watson's worries were right on the mark about her husband's financial acumen. In early 1916 Tolman put virtually all of his money at risk through betting on highly speculative penny stocks. By the summer's start, they lost most of their value. Then insufficient family monies existed to pay Dad's tuition and board charges if he had been allowed, as was very likely, to be readmitted to Oberlin. If Nellie had not kept her inheritance under her control, they would then have been completely destitute. In order to let Dad's brother Bill—academically stronger and one year younger—enter the University of Chicago (U of C) to pursue a career in science, his mother sold her house in La Grange and moved directly east to a multibedroom apartment on Harper Avenue in Hyde Park, on Chicago's South Side.[4] From there Bill could conveniently walk to the U of C. To prepare himself for a business career, Dad, at his mother's urging, became proficient as a stenographer, easily finding a job in the Loop at the Harris Trust Bank. The monies so earned let him in part make up for the huge hole in family income that followed his father's penny stock misadventures.

A Mabel Sykes portrait of James D. Watson, Sr., in 1917.

[4] James in his letters home from WWI also supported this plan. "Bill has probably graduated by this time and is working.... I know he came through with flying colors. If he needs any money in the fall let him use what I send home." —JDW, Sr., June 19, 1918, somewhere in France

Tolman in later years.

James D. Watson, Sr., in his World War I uniform.

[5] "As a whole we are not a fighting regiment. About all the men can do is to drill and handle a rifle and our company cannot even do that. Since I have been here I have not learned a darn thing about anything. Have been detailed as a Stenog. at Regimental Headquarters and have not had time for anything else." —JDW, Sr., April 20, 1918, Camp Logan, Texas

Quickly missing the intellectual challenges of his past Oberlin life, Dad saw the United States' April 1917 entry into World War I as a much-welcomed opportunity to widen his horizons.[5-14] Enlisting, rather than waiting to be drafted after the United States had begun mobilizing to send troops to France, Dad's initial day in the Army began on a train taking him to Camp Logan in Houston, Texas.

[6] "I am sorry to learn that Tom has gone to sea but it had to come…. Let me know if you learn his address. Do not neglect to send me some papers, magazines, or any little thing. It keeps me more cheerful to hear from home often." —JDW, Sr., May 1, 1918, Camp Merritt, New Jersey

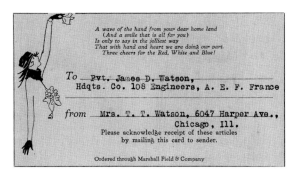

A Marshall Field & Company card from Nellie, presumably accompanying supplies requested by JDW, Sr.

[7] "I bet old Jackson Park is looking beautiful now—I wish I could take a few more bird trips, but more serious work calls me now. My idea of Heaven after this way is a nice little lake up north with plenty of eats and nothing to do but eat and sleep." —JDW, Sr., May 3, 1918, Camp Merritt, New Jersey

[8] "My bed consists of a pile of straw and a few blankets thrown over to keep out the spiders and ants. I have forgotten how it feels to sleep in a real bed. We have many luxuries and pleasures to look forward to when we come marching home after this mess is settled to the satisfaction of the men responsible."—JDW, Sr., May 28, 1918, somewhere in France

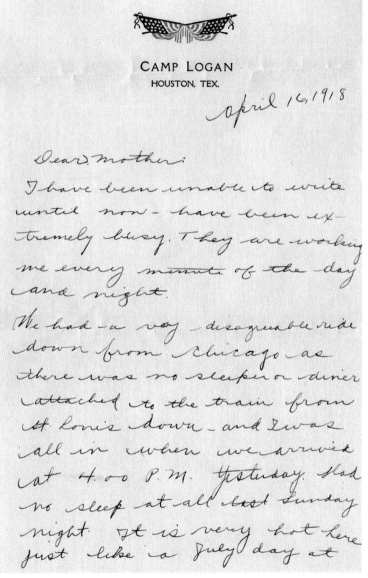

An April 1918 letter to his mother, Nellie, from Camp Logan in Texas. "We had a very disagreeable ride down from Chicago as there was no sleeper or diner attached to the train…."

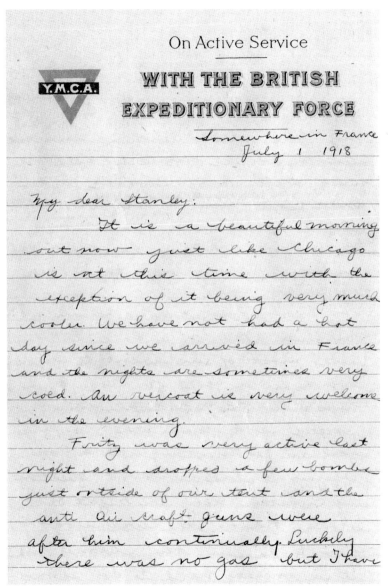

A "Somewhere in France" letter to JDW, Sr.'s brother Stanley, reporting nearby action by German planes.

[9] "I have had no time to read or study or such. Work eat and sleep is all a soldier does over here and very often the last two are missing. I am glad that Bill sent me some magazines—I do not know whether they will come thru or not. No packages are allowed to be sent over here. I do not know about printed matter." —JDW, Sr., June 18, 1918, somewhere in France

[10] "Fritz seems to be taking a rest altho he did send a few shells near me the other day. I was near enuf to an 'arrivé' to pick up a small, jagged portion of it where it had fallen on the road." —JDW, Sr., July, 1918, somewhere in France

[11] "I did not get a chance to go out very much or see much of the actual war work. Sometimes at night I go over to a village not far distant where I '*parlez vous*' with the natives. But sad to say my French is rather limited as I have no chance to acquire it thru association with the people." — JDW, Sr., July 14, 1918, somewhere in France

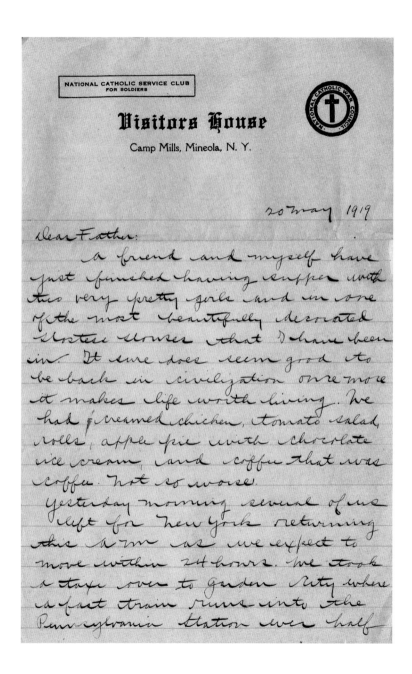

Back from the war in Europe, JDW, Sr., writes to his father from Mineola, New York, in May of 1919, describing a visit to Manhattan, including a show at the Winter Garden, and a stop at "the Jamaica race track."

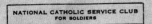

NATIONAL CATHOLIC SERVICE CLUB
FOR SOLDIERS

Visitors House

Camp Mills, Mineola, N. Y.

hour! Exceptionally good service.
After lunch we took one of
the 5th Ave busses and went
up to 150th st. and back to
the battery. We passed Central
Park and most of the big buildings. Fifth Ave is one of the
most wonderful streets I have
ever been on. The shops are
beyond description. Later we
went up to the 57th st YMCA
and had a swim in their big
pool. At night we took in
a show at the Winter garden.
This morning coming back we
stopped off at the famous race
track and looked things over.
We made every hour of the
24 count.
    I have just finished looking
over Sunday's Tribune, and notice

12 "I was walking from the office
to my billet this morning when
a Fritz shell came buzzing over
and burst 250 yards in front of
me. Believe me I made the distance between where I was and
the billet in nothing flat, just
before the second one landed,
somewhat nearer." —JDW, Sr.,
September 22, 1918, somewhere
in France

JDW, Sr., letter. (Figure continued on following page.)

13 "I have only heard from candidate W.W. Watson once since he went down to Fort Monroe, but, of course, he is very busy and had little time to write.... I know he will make good." —JDW, Sr., November 28, 1918, Tully sur Meuse, France

14 "The United States rapid but sure rise to be the World's greatest Power, had been nothing short of miraculous. Let us hope that it is not of the mushroom variety. One thing I am very sure of, and that is the more I see of the European races, the more proud I am to be an honest-to-god American. The Yank has been, is, and will be for a long time the best there is." —JDW, Sr., February 19, 1919, Echternach, Luxembourg

JDW, Sr., letter continued from previous page.

CHAPTER 3

The Kirtland's Warbler
(1920–1924)

JDW, Sr.

Upon his return from wartime service in France, Dad found a well-paying job at LaSalle Extension University (4100 Michigan Avenue), only two miles distant from his parents' apartment on Dorchester Street in Hyde Park. This highly regarded "correspondence school," in which school lessons were sent to students by mail, provided instruction in business subjects like accounting and the law. Students then as still now need not obtain law school degrees to become practicing attorneys. To become one, all that is required is to pass the bar exam in your respective state. Dad was hired to collect money from students who were delinquent in their monthly payments. His well-crafted letters propelled him into later becoming LaSalle's collection manager.[1]

Less than a mile from where Dad lived with his parents and three brothers was Chicago's largest public space—Jackson Park—sited on landfill dumped into Lake Michigan to create the sculptured Venetian-like lagoons of the 1893 World's Columbian Exposition. Between two of these broad bodies of water was the "Wooded Island," then emptied of all its former White City Pavilions, except for an increasingly run-down, wood clad Japanese tea house. During the spring and fall, flock after flock of migrating birds following Lake Michigan's western shore would preferentially descend upon the greenery of Jackson Park, most often to the Wooded Island. There its not yet tall trees and low shrubs provided ideal viewing platforms for the more than 20 different species of American warblers that annually migrate to and from their nesting sites to the north

[1] Adlai E. Stevenson, former Vice President of the United States and grandfather of the presidential candidate Adlai Stevenson, was one of the founders of LaSalle Extension University.

Jackson Park's Wooded Island, here at the 1893 World's Columbian Exposition, which ran from May through October of 1893 and celebrated the 400th anniversary of Columbus's arrival in the New World. The island sat in a lagoon surrounded by the 200 or so Exposition buildings. Designed by the famed landscape architects Frederick Law Olmsted and Calvert Vaux, the island is a bird-watching favorite, with over 250 species nesting on or passing through the island.

[2] **Frederick Law Olmsted and Calvert Vaux were originally hired to plan Jackson Park—first known as South Park—in 1869. They were brought back to Chicago in 1890 when Jackson Park was chosen as the site of the Exposition.**

of Chicago. In choosing to live so near to Jackson Park, his parents had let Dad almost effortlessly hone his skills as a seasoned ornithologist.[2]

Birding works best when several pairs of trained eyes and ears simultaneously traverse good birding sites. When a seldom seen species, say, the more southerly nesting Worm Eating Warbler, is spotted, the fact that two experts concur makes the identification much more convincing. Dad's postwar birding life thus became much enhanced upon his chance meeting in Jackson Park, soon after he came back from France, with the highly intelligent, 15-year-old birder Nathan F. Leopold (known as "Babe" by his family and close friends). Nathan's skill with his shotgun was to eventually let him collect some 3,000 bird skins that he stored in his family's large house on Greenwood Avenue. The City of Chicago had given him permission to shoot birds in its public parks. Nathan's fascination with birds dated back to when he was only five, with his later being much encouraged to do serious ornithology by teachers at the Harvard Preparatory School near the Leopold home in posh Kenwood, an enclave

The Chicago home of the Leopold family at 4754 Greenwood Avenue.

of wealthy, largely German-Jewish American families located just a mile north and west from where Dad's family lived much more modestly.

Nathan's family's great wealth had its origins when Nathan's grandfather, Samuel F. Leopold, emigrated from Germany in 1846 to establish a series of retail stores in the just discovered, rich copper belt of Michigan's Upper Peninsula. Then all the objects he sold from his store had to come by boat as train tracks had not yet been extended that far north. In 1863, Samuel farsightedly bought his first freight steamship, adding several more over the next decade to become owner of the Lake Michigan and Lake Superior Transportation Company. By the time of his death in 1899, having moved more than two decades earlier with his wife and six children to Chicago, he owned the largest shipping line in the Great Lakes. His eldest son, Nathan Freudenthal Leopold, Sr., in turn, amassed an equally large fortune manufacturing paper bags and cartons. At the time of the birth of his youngest son, Nathan F., Jr., on November 19, 1904, he was not only one of the wealthiest citizens of Chicago but well

[3] Eggers Woods, which figures more sinisterly later in this chapter, is located in the Calumet Open Space reserve and contains savanna, woodland, and marshland habitats ideal for viewing a wide range of bird species.

connected to Chicago's financial elite through his marriage to a daughter of the highly prosperous banker Gerhardt Foreman.

Like my father, Leopold and his Harvard School fellow obsessive birder, George Porter Lewis, kept meticulous records of their bird sightings, especially focusing on the first day in the spring when specific migrating species reached the Chicago region. In 1920, the three of them collectively assembled their records into an 18-page pamphlet entitled *Spring Migration Notes of the Chicago Area.* Key to many of their rare bird sightings was the availability of the several large touring cars owned by the Leopold family, all maintained in tip-top shape by the family chauffeur. Their use allowed them to spot the migrating shorebirds that descended into the marshy mudflats of Calumet and Wolf Lakes to the south of Jackson Park and to the oft-waterlogged trees and shrubs of nearby Eggers Woods Preserve on the Illinois–Indiana borderline. Use of a Leopold car let both the bird-rich marsh regions of Chicago's southern borders and Jackson Park be visited on the same day.[3]

A year after he and Dad started watching birds together, the intellectually precocious Nathan entered the University of Chicago (U of C) in the fall of 1920. Then he was 15, one of the youngest students the U of C had ever enrolled. In part due to social unease with his fellow freshmen students, Nathan transferred for his sophomore year to the University of Michigan. There he could remain near his very close Kenwood friend, the highly gregarious Richard (Dickie) Albert Loeb born on June 11, 1905. His family had even greater wealth than the Leopolds, his father Albert Loeb being second in command to Julius Rosenwald at Sears Roebuck, whose large department stores were Chicago's largest employer. Richard unwisely had entered the U of C at the even earlier age of 14, but after two years going nowhere, he decided to move to the much less intellectual life of Ann Arbor where he knew boys he had grown up with in Chicago.

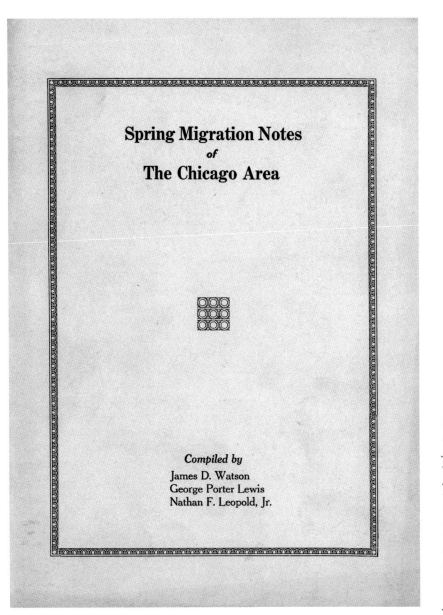

Spring Migration Notes
of
The Chicago Area

Compiled by
James D. Watson
George Porter Lewis
Nathan F. Leopold, Jr.

Cover, Introduction, and two pages from the compilation of bird species sighted from 1913 to 1920 by Watson, Lewis, and Leopold. This privately published pamphlet was reviewed in a 1920 issue of The Auk, *in which the reviewer commented that, "The list seems to be very carefully prepared and should be of much interest to other bird students of the district, while it will also furnish a convenient comparative record for those interested in the general study of bird migration." (Figure continued on following pages.)*

Introduction

THE following list of migration dates has been compiled entirely from the lists of Messrs. Locke L. Mackenzie, Sydney Stein and ourselves. These lists cover the spring migration for the eight years 1913-1920. and include the earliest date for each bird each year.

All specimens were either taken or clearly seen, and the list should contain some valuable dates. No bird, not actually seen by one of the above mentioned has been included.

By the words Chicago Area, we mean that portion of the country lying within a radius of thirty miles from the city itself. This region includes several excellent stations for observation, such as Miller, ,Ind., Hinsdale, Argo, Riverside, Highland Park, and Hyde Lake, Ill., and the city parks, a diversified region containing lakes, woods, swamps, and rivers. The list contains two hundred and twenty-seven species seen in this area, five seen only in the fall migration, and five sub-species whose occurrence in our area is doubtful. This totals two hundred and thirty-seven species and sub-species. The names used are those given in the 1910 A. O. U. check list.

We wish to express our sincere thanks and gratitude to Messrs. H. K. Coale, Locke Mackenzie, and Sydney Stein, whose kind co-operation has made the work possible.

THE AUTHORS.

Compilation of bird species continued from previous page.

SPRING MIGRATION NOTES OF THE CHICAGO AREA   (3)

	1913	1914	1915	1916	1917	1918	1919	1920
2. Holboell's Grebe				Mar. 20		*Colymbus nolboelli*		

Rare visitant; winters on Lake Michigan. One bird remained in the park lagoons several days.

	1913	1914	1915	1916	1917	1918	1919	1920
3. Horned Grebe	Apr. 13			Apr. 8	Apr. 9	Mar. 24	Apr. 5	Apr. 28

Colymbus auritus

Common migrant.

6. Pied-billed Grebe	Mar. 30	Mar. 26	Apr. 2	Mar. 25	Mar. 26	Mar. 15	Apr. 5	Mar. 30

Podilymbus podiceps

Common summer resident; most common during migrations.

7. Loon					Apr. 7	Mar. 16	Apr. 1	Apr. 7

Gavia immer

Fairly common migrant.

11. Red-throated Loon			Feb. 21					

Gavia stellata

Very rare winter visitant. One, apparently wounded observed in Jackson Park, Feb. 21, 1915. Woodruff reports three dead birds at Evanston.

51. Herring Gull	Jan. 1	Jan. 1	Jan. 1	Jan. 2	Jan. 1	Jan. 1	Jan. 1	Jan. 1

Larus argentatus

Abundant winter resident: a few non-breeding birds remain during the summer.

54. Ring-billed Gull	Apr. 26	Jan. 2	Jan. 11	Jan. 2	Apr. 6	Jan. 18	Mar. 14	Jan. 1

Larus delawarensis

A fairly common, but erratic visitant.

60. Bonaparte's Gull	Apr. 14	Jan. 1	Mar. 29	Mar. 27	Mar. 21	Mar. 28	Jan. 1	Mar. 26

Larus philadelphia

Common migrant, sometimes wintering. A few non-breeding birds remain during the summer.

69. Forster's Tern			Apr. 4		May 2			May 3

Sterna forsteri

Uncommon migrant, but probably more common than generally supposed, especially in the fall at Miller.

70. Common Tern	Apr. 30	Apr. 25	Apr. 26	May 2	Apr. 21	Apr. 17	May 9	Apr. 20

Sterna hirundo

Common migrant.

77. Black Tern	May 24	Apr. 21	May 9		May 1	May 18	May 24	May 1

Hydrochelidon nigra surinamensis

Breeds commonly in the swamps and small lakes in the vicinity.

129. Merganser	Apr. 2	Mar. 9	Jan.18	Jan. 2	Feb.19	Jan.26	Jan. 1	Jan. 1

Mergus americanus

Common winter resident.

130. Red-breasted Merganser	Mar.30	Mar.30	Mar.25	Jan.23	Mar.31	Apr.13	Mar.18	Mar.18

Mergus serrator

Abundant migrant.

131. Hooded Merganser		Apr. 21	Mar. 1				Apr. 5	Mar. 26

Lophodytes cucullatus

Rare summer resident.

SPRING MIGRATION NOTES OF THE CHICAGO AREA   (5)

	1913	1914	1915	1916	1917	1918	1919	1920

165. White-winged Scoter   Oidemia deglandi

    Apr. 18

Uncommon winter visitant.

167. Ruddy Duck   Erismatura jamaicensis

    Apr. 15   Apr. 16   Apr. 9   Apr. 17

Not uncommon migrant.

171a. White-fronted Goose   Anser albifrons gambeli

    Apr. 7

Rare migrant. A small flock seen in Hinsdale swamp by Mr. Watson.

172. Canada Goose   Branta canadensis canadensis

Feb. 23   Jan. 25   Mar. 8   Jan. 3   Mar. 10   Feb. 10   Mar. 25   Mar. 3

Common migrant often wintering; spends the day on the edge of the ice on the lake, and the night inland on some cornfield of the previous season

190. Bittern   Botaurus lentiginosus

    Apr. 28   Mar. 21   Apr. 11   Apr. 24   May 4   Apr. 7   Apr. 22

Common summer resident.

191. Least Bittern   Ixobrychus exilis

    May 1   Apr. 27   May 8   June 8   May 30   May 10   May 15

Fairly common summer resident.

194. Great Blue Heron   Ardea herodias herodias

    May 11   Apr. 29   Mar. 31   May 10   Apr. 3

Not uncommon summer resident.

200. Little Blue Heron   Florida caerulea

    May 8

Casual visitant. One observed in Jackson Park, May 8, 1915.

201. Green Heron   Butorides virescens virescens

Apr. 22   Apr. 15   Apr. 22   Apr. 16   Apr. 18   Apr. 28   Apr. 19   Apr. 22

Fairly common summer resident; common during migration.

202. Black-crowned Night Heron   Nycticorax nycticorax nacvius.

Mar. 29   Apr. 20   Apr. 18   Mar. 31   Apr. 28   May 10

Rather common summer resident.

203. Yellow-crowned Night Heron   Nyctanassa violacea

    Apr. 14

Accidental straggler in the area. One, observed in Jackson Park, April 14, 1918. The bird showed no signs of having been in captivity, and flew well.

206. Sandhill Crane   Grus mexicana

    May 2   Mar. 24   Apr. 22

Formerly common, but now extremely rare migrant. One seen in Jackson Park, attempting to fly against a strong wind. It continued its efforst for half an hour but was finally blown out of sight to the south.

208. King Rail.   Rallus elegans

    Apr. 19   Apr. 24   Apr. 25   May 20   May 4

Common summer resident.

212. Virginia Rail.   Rallus virginianus

Apr. 22   Apr. 25   Apr. 18   Apr. 25   Apr. 18   May 5   May 10   Apr. 24

Common summer resident.

Compilation of bird species continued from previous page.

Norman Asa Wood, who spent his entire career at the University of Michigan, discovered the first Kirtland's Warbler nest in Michigan. He studied the birds of Michigan exclusively, having visited every county in the state to map nesting sites.

After opting to tag along with Dickie, Nathan first had to cope with the loss of his beloved mother, Florence, to a long fatal illness in 1921. Soon after he also lost the day-to-day companionship of Dickie, whose newly joined Jewish fraternity Zeta Beta Tau vetoed Nathan as a member. They did not like the persistent rumors of the two boys' friendship being closer than they were prepared to tolerate. The only saving aspect of Nathan's Michigan year was a birding internship under Norman Asa Wood (1857–1943) of the University of Michigan Museum. He was famous for his 1903 investigation of the breeding range of one of the rarest of American birds, the Kirtland's Warbler.[4]

[4] Wood described an early encounter with the Kirtland's Warbler in *The Jack-Pine Warbler*—"As I sat near the nest, the female came and alighted on the branch of the jack pine just back of the nest…. The male also came, but not so close."

The Kirtland's Warbler, Dendroica kirtlandii. *This sparrow-sized songbird is listed as an endangered species under the Endangered Species Act.*

To see it firsthand, Nathan enlisted Dad to join him on his June 1922 journey to the Jack Pine Region of Lower Michigan. After a long 300-plus-mile drive north to Traverse City, they went 75 further miles eastward on almost impassable dirt roads to the tiny town of Luzerne near the banks of the Au Sable River. There Norman Wood had secured specimens in 1903. The trip, however, yielded no Kirtland sightings, and they wondered whether the growth of its Jack pines over the past 20 years somehow made it a less likely home for the birds. Before returning to Chicago they made a brief detour north to Charlevoix to see Dickie Loeb, then vacationing with his parents at their palatial summer home overlooking Lake Michigan.[5]

When returning the next year back to the U of C, Nathan was much more at ease with his fellow students and through working very hard graduated, at the age of 18, in March of 1923 with a Phi Beta Kappa key after less than three years of instruction. It was obvious to all that he had become the U of C's most brilliant student. After the spring migration had passed and the summer nesting season had commenced, he and Dad started their second trip back to the Jack Pine Region on June 16. Encouraging them to make the second trip was Nathan's learning from Norman Wood of several new Kirtland sightings the previous summer along the Jack pine belt, some 145 miles in breadth that stretched across the state of Michigan from Traverse City to Oscoda. To magnify their chances for success, they had along with them their young Chicago birding friends, Sydney Stein, Jr. and Henry B. Steele.

Kirtland spotting came quickly with first Sydney hearing an unfamiliar song and then my father and Nathan spotting the singer in full nuptial plumage in a large pine tree. The next day they located their first Kirtland nest on the ground in a thick clump of Jack pines. A chance lunchtime meeting in Oscoda with James MacGillivray, then a member of the Michigan State Conservation Department, led the next day to his making motion pictures of Kirtland's Warblers and their nests.[6]

[5] Loeb graduated from the University of Michigan at age 17, then the youngest graduate of the University. He enrolled in a constitutional history course at the U of C and renewed his friendship with Leopold, who was then studying at the University of Chicago Law School.

[6] James MacGillivray was a newspaperman who turned to conservation work in 1911 with the Michigan State Conservation Department. He is also known for popularizing the tale of Paul Bunyan, first in a 1906 issue of his brother Will's local newspaper the *Oscoda Press* and then in his article, "The Round River Drive," in a 1910 issue of the *Detroit News Tribune*.

Stills from the MacGillivray film, The Kirtland Warbler in Its Summer Home, *showing James D. Watson, Sr. (in sweater and tie) and other members of the expedition. The film was presented at the 41st meeting of the American Ornithologist's Union in October of 1923.*

Nathan's skill as a hunter then let him add several Kirtland's Warblers to his collection. This made Dad uncomfortable, fearing that the loss of only a few birds might tip this species toward total extinction. Happily their range today extends across Lake Michigan to the Jack Pine

Nathan Leopold with a very tame Kirtland's Warbler.

Region of northern Wisconsin and into Canada. Several of the birds then shot long graced a display case on the Kirtland's Warbler at the Cranbrook Institute in Detroit, while the Field Museum's (in Chicago) Kirtland's Warbler skin still bears Leopold's name on it. Also among its collections are several shorebird skins from the "Boys' Club of Chicago" thought to be from Leopold as they were collected around Hyde Lake across the Pennsylvania Railroad tracks from the Eggers Wood Preserve.

Over the summer Nathan emerged as a visible new force in American ornithology through his paper "Reason and Instinct in Bird Migration," which appeared in the July issue of *The Auk* (40: 409–414), the premier ornithological journal of the United States. By summer's end, he had completed a second article, "The Kirtland's Warbler in Its Summer Home," illustrated by stills from James MacGillivray's movie that showed Nathan feeding horseflies to a male parent. In October, he delivered the paper in person to the Annual Meeting of the American Or-

Vol. XL]
1923]   LEOPOLD, *Reason and Instinct in Migration.*   409

tent by the excited cries of the old male who was trying to drive off a persistant young one who had appeared on his premises and who had to be chased out of the tree again and again. Then two fresh-looking cup-shaped nests of the shallow Hooded type were discovered in the leafy sycamores opposite the ranch house, and the birds were seen flying busily around.

The mesquites which had now come into fresh green leaf were fragrant with yellow tassels. And here and there big cactus flowers were to be seen. The desert was putting on bridal garments.

1834 Kalorama Round, Washington, D. C.

REASON AND INSTINCT IN BIRD MIGRATION

BY N. F. LEOPOLD, JR.

IT IS a well known fact that every species of animal, having had its origin in a very small section or territory, tends to spread farther and farther in all directions from its center of origin. This gradual spread continues until progress is checked by a barrier of some variety. This entire principle has been summed up under the zoological law of "barrier control dispersal from a geographic center of origin."

It is very easy to explain this tendency, and the resultant gradual extension of range of resident animals, but when we attempt to carry our explanation further, to the extension of migration ranges in birds for example, we are confronted with a more difficult problem.

In the case of resident animals it can easily be understood why this tendency must exist. A species springs up in a limited area; its numbers increase; the food supply in the original area proves inadequate to satisfy the needs of the increasing number of individuals and some of the more hardy members of the race go forth to seek new territory and a new food supply. Their selection of direction followed in this quest for food is a matter of chance, but those going in a direction where they find conditions favorable, survive, and settle there permanently; while those which are less fortunate in their choice of new quarters are either killed off, or

Leopold's precocious paper on bird migration, published in The Auk, *explored the roles of instinct and learning in migration and was illustrated with observations on birds in the Chicago area made by Leopold and others, including James D. Watson, Sr. (Figure continued on following pages.)*

forced to move to still other territory. Such a tendency to dispersal seems logical enough—in fact is absolutely necessitated by food conditions, but now let us turn our attention to the problem of bird migration.

An explanation of the whole problem of bird migration was once attempted upon the same line of reasoning as pertains in the case of non-migratory animals. For example, a bird nesting perhaps in the extreme northern portion of the Temperate Zone of the Northern Hemisphere found conditions in this locality favorable during the summer months, but when winter set in, it was forced to seek new territory where climatic conditions were less rigorous, and the food supply more abundant. Here again the selection or choice of which course to follow, north or south, was a chance matter, but those who chose the former perished; those who chose the latter survived. This same problem, so ran the old theory, presented itself each year, and each generation was compelled to make its own choice or perish. This belief of course is controverted by the fact that most birds leave their breeding ground long before climatic conditions render it necessary, and besides this, that many birds traverse a much longer route in their semiannual migrations than would be necessary to find suitable climatic conditions. A case in point is the Golden Plover (*Pluvialis dominicus*) which leaves its breeding ground in the Arctic circle in July, the most clement part of the year, and in its migration south, passes over the Temperate and Semi-tropic Zones of the Northern Hemisphere, where favorable conditions exist in order to continue its long flight to Patagonia.

These two facts absolutely preclude the possibility of the suggestion that the question of bird migration is a matter decided by each generation of birds for itself, and nearly all zoologists today hold the belief that a tendency toward migration is inherited by each generation of birds from its forebears as an instinct. This accounts for the fact that birds leave their summer homes long before climatic conditions compel them, "instinctively" knowing that such conditions will come. The fact that birds ignore favorable localities in their migrations, in order to go farther where similar conditions exist, is explicable by the same line of reasoning. The instinct which causes birds to migrate, originated

*Leopold's migration paper continued
from previous page.*

Vol. XL]
1923] LEOPOLD, *Reason and Instinct in Migration.* **411**

at a time when the necessary conditions of food supply and climate could not be duplicated in any region nearer to the breeding grounds than that chosen by the species for its winter home. But conditions have changed. Today the Golden Plover could find in our own Southern States conditions identical with those in his winter retreat. Why then does he not shorten his migration route by 1,000 or 1,500 miles and take up his winter abode in the Northern Hemisphere? Will he ever change his habits in this way? This is the question which I wish to attack in this paper.

Before attempting to explain any mode of behavior on the basis of instinct, it is of course necessary for us to procure a definition of the term instinct. In most modern schools of psychology, instinct is defined as an inherited pattern of behavior; the inheritance of a habit once acquired by the ancestors of the animal in question. The definition would do very nicely were it not for the fact that evolutionists and students of heredity have proved to us, without question, that an acquired trait, and one which effects only the body or somatic cells without influencing the germ plasm is incapable of being transmitted to the offspring through heredity. Such an acquired trait a habit obviously is, and hence it becomes impossible to adhere to the definition of an instinct as an inherited habit. A small number of psychologists recognizing the absurdity of the old definition have substituted therefore the following: "An instinct is an inherited or native tendency toward a particular method of behavior." This, while ambiguous and not wholly satisfactory, would do very well as a working definition in the present problem. Now let us look into the matter. Is instinct the only factor in migration?

If instinct is the one and only factor governing bird migration then the Golden Plover which instinctively turns to Argentine and Brazil as its winter home, will never give up this locality in favor of equally favorable and nearer areas. Reason and learning cannot enter at all. The birds' habits are instinctive and will remain fixed until natural conditions require their change.

But is this the case? Let us take as an example the case of two birds and their status in the Chicago Area. This area is situated ideally for a study such as this. At the southwestern tip of Lake Michigan, east but not too far east of the Mississippi

Valley, Chicago offers a wonderful place for the study of accidental visitors and stragglers from the west. Its position on the Great Lakes makes it a logical place to which will be blown birds carried from their normal habitat by storms.

The first bird I shall quote as an example is Harris's Sparrow (*Zonotrichia querula*) a distinctly western species, common in migration west of the Mississippi River. Until very recently the bird, a rare, almost accidental visitant in the Chicago area, was included on our list on the strength of two positive records and two sight records. In the last three years however there have been at least twelve records, two of Mr. H. L. Stoddard at Miller, Indiana; three by Mr. Lyons at Waukegan, Illinois; one by Mr. Sanborn at Beach, Illinois, and six by Mr. G. P. Lewis and myself at Chicago, Illinois. But more than this, the vast majority if not all these records were made in the fall, all between September 20 and October 1 each year. Still further, the bird has not skipped a single one of the last three years, and lastly, several of the specimens, at least one of Mr. Lyons' and my last specimen (taken September 26, 1922) were immature birds. How shall we account for this phenomenon; a bird accidental at Chicago for the last sixty years, suddenly grows regular and commoner in its occurrence. I should like to suggest the following explanation.

The first individuals appearing here, appeared from accidental causes. Perhaps they were blown out of their course by a storm, and found shelter here. Upon finding themselves in new territory, they learned that here, in a place foreign to their normal migration route, conditions were favorable to them. The next year instead of following the traditional path, far to the west, they again chose the short cut upon which they had stumbled the year before, —and they brought their brood with them. This is not instinct, it is *reason*. An objection may be urged against this theory on the ground of coincidence. This would hardly explain how three successive years, the birds have chosen almost exactly the same date, and have chosen exactly the same field in which to put in in their appearance as is the case with the records of Mr. Lewis and myself. The birds chose for the first two years a small field in a rather well populated district of the city, just south of the Jackson Shore Apartments and north of Jackson Park. Apparently

Leopold's migration paper continued from previous page.

Vol. XL]
1923]   LEOPOLD, *Reason and Instinct in Migration.*   413

a certain weed upon which the bird feeds was available here. This last year however, the field has been cleared of vegetation, and the bird has sought a new habitat, alighting upon similar fields at the northern extremity of Lincoln Park, a distance of about eleven miles from its former haunts. Another case in point is the Franklin's Gull (*Larus franklinii*). Once before reported from our area, and only twice from the State of Illinois, this bird was noticed on October 15, 1921 on a small sand-bar at the North end of Lincoln Park by Messrs. Lewis and Watson. At the time of discovery there were about ten birds in the flock and hence it would be necessary to assume either that the whole flock was blown in by a storm, or that the discovery marked a rather advanced stage in the Gull's extension of migration route, and that previously straggling individuals had found their way here but had not been noticed. This would be an easy and natural explanation in view of the similarity of this bird to the common *Larus philadelphia*, and the consequent difficulty of distinction. Furthermore the birds were again noticed on the identical sandbar and on almost identically the same date (October 23, 1922) and the birds acted in a precisely similar manner to last year;— gradually decreasing in numbers until about a week after their first discovery they had all left. In this case also a large proportion of the birds were immatures, the brood perhaps, of those pioneer birds which the year before had either accidentally or purposely selected a migration route equally favorable with the old one; definitely controverted their old "instincts;" and by reason repeated their trip of last year.

There are numerous other examples of birds extending their range into our area, although the two cited constitute most striking examples. The Western Meadowlark (*Sturnella neglecta*), and the Bachman's Sparrow (*Peucea aestivalis bachmanii*) are examples of summer residents where this extension of range has occurred. The former, a distinctly trans-Mississippian form was unknown in the Chicago area until recently when it has established itself firmly in several nesting colonies, one in Chicago Heights, and one near Rockford, Illinois. The latter, a southern form has only recently been found to nest at River Forest, near Chicago where its numbers increase each year.

Leopold's migration paper continued from previous page.

414　　　De Lury, *Migration in Relation to Sun-spots.*　[Auk [July

As permanent residents the Cardinal (*Cardinalis cardinalis*) and the Tufted Titmouse (*Baeolophus bicolor*) may be cited. Both southern forms common in the southern portion of the State, have persistently pushed the boundaries of their range northward, until today the once rare Cardinal is very common if not abundant, and the formerly accidental Tufted Titmouse is a not uncommon permanent resident in restricted parts of the area. These cases however are not examples of changes in migration route, and hence are not so important to our subject as the first mentioned.

Now then to sum up. If we believe that instinct is the only factor influencing bird migration, how can we account for such instances as the above? Shall we not rather say that instinct without doubt is the motivating impulse in the idea of migration in general, but that the change of specific migration route, though brought about by chance, is in many cases preserved and continued by reason of learning, until it again appears in later generations as instinct.

4754 Greenwood Ave., Chicago, Ill.

[7] Leopold notes in his Kirtland's Warbler paper in *The Auk* that "Mr. Watson was walking through the pine attempting to get a look at the bird, he started a female from the ground about two feet from where he was standing. The bird half flew and half hopped a few feet, and then hopped slowly into the shrubbery, dragging one wing as though it were broken. This so-called broken wing ruse is, of course, common enough among many other birds, notably the Meadowlark, but to my knowledge has never been observed before in the case of the Kirtland's Warbler."

nithologists' Union. Three months later, it came into print in the January issue of *The Auk* (41: 44–58).[7]

During the fall quarter of 1923 Nathan was taking four courses, one more than usual, now at the U of C Law School from which he hoped to transfer to Harvard Law School the following school year, when he would finally be old enough to be admitted to its first-year classes. For the first time he was acquiring friends who sought out his higher intelligence despite his all too obvious sense of superiority. Weekly he invited them to his home for discussions of possible exam questions and important law cases. To facilitate their flow, Leopold typed up study sheets, making carbon copies for all his fellow participants. Then, also eager to exploit his wide-ranging linguistic skills, he and his high-stakes bridge partner, the department store heir, Leo Mandel, began translating into English the 16th century narrative *I Ragionamenti* by the Italian satirist Aretino (1492–1556). Their efforts abruptly halted when the head of

THE KIRTLAND'S WARBLER IN ITS SUMMER HOME.

BY N. F. LEOPOLD, JR.

Plates VII-VIII.

DURING the spring of 1922, it was my privilege to work under Mr. Norman A. Wood of the University of Michigan Museum, the discoverer of the breeding range of Kirtland's Warbler, and at his suggestion I determined to make a trip to the region where he had discovered the type nest and egg of the species. Consequently on June 24, 1922, in company with Mr. James D. Watson of Chicago, I started in my automobile bound for Luzerne, Oscoda County, Michigan.

The first night was spent at Pentwater, Michigan, at a distance of about 280 miles from Chicago, and at noon of June 25, we arrived at Traverse City. From this point it was necessary to strike due east across the state over well-nigh impassable roads, and as a result we were compelled to spend the night at Grayling, still at some distance from our destination, but at least in the same character of country in which we expected to search for our bird. Needless to say we were constantly on the lookout for unfamiliar songs and birds, but although we traversed many suitable stretches

Opening page of Leopold's very detailed report in The Auk *of the June 1923 Kirtland's Warbler expedition.*

the Italian language department heard of the project that described human behavior inappropriate for print. He promptly informed Mandel's parents, who quickly sent him to Europe to remove him from Leopold's influence.

Nathan's father by then also saw his son going dangerously out of control, his own worries coming from the excessive evening hours Nathan was spending with Dickie Loeb. They were precluding any normal contact with his family. Inherently troubling was Nathan's obsessive preoccupation with the German philosopher Friedrich Wilhelm Nietzsche, who saw the goal of human evolution to be the emergence of a dominating "Superman" (Übermensch) not subject to the laws governing ordinary human beings. In Loeb, Nathan felt he had finally seen one—handsome, irresistible to girls, and highly intelligent. Nathan's father, thinking otherwise, contacted my father by mid-fall to ask him to try to steer his son away from Loeb, back into more birding. But with the

Title page of Pietro Aretino's book published ca. 1534–1536. It scandalously included dialogs with prostitutes as well as with nuns and married women.

[8] In his book *Life Plus 99 Years*, Leopold says that "I thought more of Dick than of all the rest of my friends put together. His charm was magnetic—maybe mesmeric is the better word."

[9] Clarence Darrow later described Leopold's preoccupation with Nietzsche—"Here is a boy who by day and night, in season and out, was talking of the superman, owing no obligations to anyone; whatever gave him pleasure he should do, believing it just as another man might believe a religion or any philosophical theory."

fall migration effectively over, Dad had no way to move soon again into Nathan's ever-multifaceted orbit.[8,9]

When the spring migration of 1924 resumed, Leopold birded largely with his more similarly aged friends, George Porter Lewis and Sydney Stein. Nathan also was leading paying field trips for aspiring young ornithologists at the Harvard School and at the University of Chicago High School. When Nathan's schedule kept him from so teaching, George Lewis would step in at short notice. On Saturday morning, May 17, looking for shorebirds in the great marshes south of Chicago, Nathan and George saw three Wilson's Phalaropes fly out of the swamp adjacent to Eggers Woods and into Hyde Lake. Failing then to bag even one of them with Leopold's shotgun, they went back the next day with Sydney Stein hoping for another chance. But the three of them saw no trace of the rarely spotted Wilson's still then thought to be occasionally breeding around Calumet Lake.

Fourteen-year-old Bobby Franks, a cousin of Dickie Loeb. In his statement to the State's Attorney on May 31, 1924, Leopold said their planned victim "was left undecided until the day we decided to take the most likely looking subject that came our way."

Only four days later, on Thursday, May 22, Eggers Woods became front-page afternoon news when a boy's body was found in a drainage culvert running from it under the Pennsylvania Railroad tracks to Hyde Lake. Fears arose and were soon confirmed that the body belonged to the Kenwood-dwelling Bobby Franks, a 14-year-old student at the Harvard Preparatory School. He had been kidnapped the afternoon before when coming home from school by two men in a large touring car. After killing him, they had stripped off his clothes before hiding the corpse from likely view by passersby on foot, and they also poured hydrochloric acid over Franks' face and body to delay identification. Not suspecting the body would be discovered so soon, the killers had telephoned Bobby's home the evening of the kidnapping saying a note would arrive the next morning telling his father, Jacob Franks, how to pay out the $10,000 ransom for releasing his son.

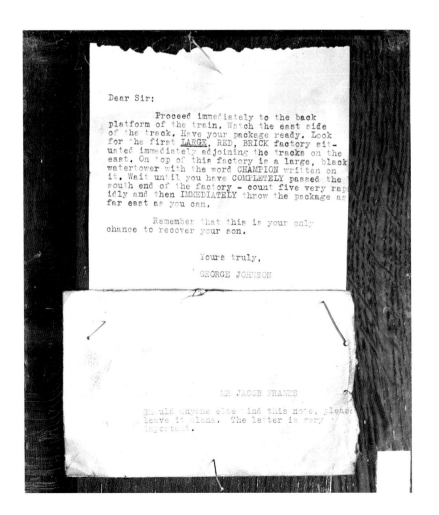

The ransom note typed on Leopold's Underwood typewriter, giving instructions to Jacob Franks for paying the ransom. Leopold and Loeb destroyed the typewriter by prying off the keys and throwing them off a bridge in Jackson Park according to Loeb's confession. The typewriter itself was thrown off another bridge in the park.

[10] Susan Lurie was one of Leopold's wealthy friends and a senior at the U of C. She described Leopold as "an interesting friend, an excellent conversationalist, and a delightful dancer" in her interview with *The New York Times*.

But before Bobby Franks' father was in a position to follow the kidnappers' instructions for delivering the ransom made up of well-worn 20 and 50 dollar bills, he learned that his son's body was the one found dead in the drainage culvert and so broke off further contact. Most unanticipated was the well-crafted, almost legalese, wording of the typed ransom note that the police soon identified as coming from a portable Underwood. The only clue so far that might actually lead to the killers was a pair of horn-rimmed reading glasses whose tips were chewed, indicating nervousness of the owner. Through questioning the game warden at Eggers Woods, the police learned about Nathan Leopold's frequent birding excursions. Early on the Sunday (May 25) following the crime, two policemen went to the Leopold home and asked Nathan to come to their police station. There he wrote out an affidavit stating that he had last visited Eggers Woods on the Saturday and Sunday before the murder in search of migrating shorebirds. George Lewis was with him on both days with Sydney Stein also present on Sunday. Then much anticipating lunch with his friend Susan Lurie, he drove to her North Side home to take her to lunch at a small country inn and afterward canoe on the Des Plaines River.[10]

Though their first police visit with Nathan aroused no suspicions, by the time of their second visit four days later on May 29, he was the prime murder suspect. Just hours before he had been identified as a likely owner of the prescription eyeglasses found near Bobby Franks's body. Upon initially claiming that he must have lost them in his home, he was taken to a high-floored room in the LaSalle Hotel, where District Attorney Robert Crowe showed him the murder scene glasses, asking whether they were his. Upon his denying it, he was taken back to his home for further searching. But when again his lookings about were to no avail, his older brother Michael brought up the possibility that Nathan must have lost the seldom-used reading glasses when birding. Since he only used them for fine print, he had no reason to notice their

Susan Lurie with State's Attorney Robert E. Crowe during Leopold and Loeb's trial. As reported in the June 2, 1924 issue of The New York Times, *Leopold had joked with her about the Franks murder. She noted that, "I had no intimation of the terrible things that were whirling through his brain."*

falling out of his jacket. Upon his return to the LaSalle Hotel room for more questioning that went into the small hours of the night, Nathan, claiming first that he had no memory of his exact movements during the murder day, finally admitted it had been spent with Dickie Loeb. After lunch at Marshall Field's, they had gone in his car to Lincoln Park for birding. Then in his car they drank much hard liquor, behavior which if Dickie Loeb's father learned would make him very angry. After failing later to pick up two girls on 63rd Street, they had gone back to their respective homes.

Unknown to Nathan, the police had also brought Dickie Loeb to the LaSalle Hotel. In a separate floor of the hotel, they grilled him on his movements during the murder day, questioning him into the early morning hours, when, like Leopold, he was moved to the police station, where they kept them from direct contact with each other. The first solid clue that both boys were lying came when two newspaper reporters, learning of Leopold's small study group of law school freshmen, unearthed several study papers that were soon identified as typed on the same typewriter used for the ransom note. Though Leopold soon contradicted his housekeeper's assertion that she several weeks before had seen a portable typewriter, the police by then thought Leopold and Loeb were the killers. Certainty came when the family chauffer revealed that Leopold and Loeb could not have been driving about in his red Willys-Knight roadster on the murder afternoon. It was then being repaired in the family garage.

Loeb was the first to confess on Saturday, May 31, 1924, at 4 AM, taking more than an hour to reveal his lurid details. Nathan only admitted his guilt when he learned that his close friend had already spilt key crime details like their renting of the green or dark blue Willys-Knight touring car. Leopold's statement was made on May 31, 1924, at 4:20 AM and only differed significantly from Loeb's earlier one in who had been the actual murderer. Both pinpointed each other. If Nathan (Dickie) was the car's driver, Dickie's (Nathan's) hand had to be the one that delivered the chisel blows to Bobby's head. The murder details that finally emerged, first from their initial confessions and then from subsequent statements made to the psychiatrists who questioned them on behalf of either the prosecution or the defense, transformed the murder of Bobby Franks from a local news story into an event that would transfix not only the United States but the entire newspaper-reading world.[11]

The crime was not one of ordinary violence or greed but reflected Leopold and Loeb's wish to have the thrill of perpetuating the perfect crime

[11] Loeb in his statement said, "I am fully convinced that neither the idea nor the act would have occurred to me, had it not been for the suggestion and stimulus of Leopold." And Leopold in his statement said, "Richard placed his one hand over Robert's mouth to stifle his outcries, with his right beat him on the head several times with a chisel."

where they never would be caught despite the demanding and receiving of a ransom requiring the victim's family be wealthy. Loeb planned to spend his half in Mexico, while Leopold's equal portion would be used in France, where he already was planning to spend the summer with a friend. Bobby Franks was only chosen at the very last moment, he fitting their objective of abducting a Harvard Preparatory School student one of them already knew. As Bobby's home was across Ellis Avenue from that of the Loeb home, he had no reason to feel danger when Dickie beckoned him into the touring car on the pretense he wanted to talk about tennis rackets. From the start of their scheming that went back six months, they saw the need to kill someone like Bobby since they were only likely to entice someone into their car that they already knew.[12]

Bobby Franks's murder was the culmination of the many acts of earlier crimes, including stealing and then destroying cars, throwing bricks through store windows, and setting off false fire alarms, as well as setting several fires, that the pair committed starting during Nathan's first year at the U of C. Of the two boys, Loeb was the much more criminally motivated, needing Leopold primarily as a witness to his inherent wickedness. Leopold's obsession was never for violence; birding came much more naturally to him. Instead, his criminal streak had its origins in a homosexual relationship that started in 1920 on a long train ride taking them to Loeb's summer home. Then Leopold was the aggressor, Loeb afterward only seeing reason to continue when he needed Leopold's presence for producing real thrills from criminal acts.

Though they led separate lives for much of two years while Loeb was at the University of Michigan, their relationship resumed once Dickie came back to the U of C in the fall of 1923. Then Nathan fell strongly in love again, considering the much more handsome Loeb a Nietzsche "Superman," not bound to moral codes constraining ordinary humans. Dickie's criminal fantasies went back to when he was a young boy and an avid reader of detective stories. He wanted to move on with Nathan

[12] **Apparently Leopold had planned ahead. On April 28, 1928, he was issued a passport to travel to the British Isles, France, and Italy, leaving from New York on June 11, 1924. In an April 4, 1924 affidavit, his father, Nathan F. Leopold, Sr., gave his son "permission to travel in Europe." Leopold later said his father was giving him $3,000 for him to spend three months on the Continent.**

[14] Clarence Darrow later described Leopold—"Babe is somewhat older than Dick, and is a boy of remarkable mind, away beyond his years. He is a sort of freak in this direction, as in others; a boy without emotions, a boy obsessed of philosophy, a boy obsessed of learning, busy every minute of his life." Leopold was "the most brilliant intellect that I ever met in a boy."

to the kidnapping of a young boy or girl for whom ransom would be demanded and obtained. Nathan for his part only agreed when Loeb agreed to have sex with him again with regularity, a demand that for Loeb was sex motivated much more by crime than sex whittled down to three times every two months. During the winter quarter the details of their projected perfect crime came together, early on deciding they would hide the body at a deserted site under the drainage culvert on the western border at the Eggers Woods Reserve. Still to be worked out was how they should both participate in the actual murder. If both were liable for the death penalty, neither would have advantage in confessing first to the police.[13,14]

Immediately after their signed confession became known, Leopold and Loeb's parents contracted the already very famous, 67-year-old attorney Clarence Darrow to defend their sons in court, hopefully keeping them from being hanged on short notice. Darrow was an inspired choice for he had long believed the ultimate cause of most crimes was the un-

Leopold and Loeb's families brought in one of the most famous lawyers of the day to defend their sons. This is the appearance notice filed by Darrow with the court.

Ad for Darrow's 1922 book, Crime: Its Course and Treatment. *In this book, Darrow attributed criminal behavior to biological determinants. He noted in the book's Preface that, "The physical origin of such abnormalities of the mind as are called 'criminal' is a comparatively new idea" and that insanity, which had been treated as a moral defect, "is now universally accepted as a functional defect of the human structure in its relation to environment." The Leopold and Loeb case would provide fertile ground for Darrow to propound these views.*

[15] Another reason for pleading guilty was to block the State's Attorney from trying the boys first on one charge (they were indicted separately on murder and kidnapping for ransom charges) and then, if that did not work out, trying them on the second charge. As Leopold later recounted, Darrow said, "But, boys, we had to do it this way! There is only one legal matter, one point of strategy involved. In Illinois there are only two crimes punishable by the death penalty. You were unfortunate to commit them both."

[16] In his autobiography, Darrow explained that, "No client of mine had ever been put to death, and I felt that it would almost, if not quite, kill me if it should ever happen."

satisfactory environments of those who commit them. He also opposed the death penalty and hoped to use this case to further this opposition. After taking on the defense of these children of great privilege, he decided that Nathan and Dickie should not try to retract confessions that were made without advice of legal counsel. By so acting, their fates would be decided by a jury of Chicago citizens all likely initially to want them hung. By instead pleading guilty, a judge as opposed to a jury would decide whether they were to hang. As Darrow later noted, "We placed our fate in the hands of a trained court, thinking he would be more mindful and considerate than a jury." Darrow's aim was to present mitigating evidence that could lead the judge not to sentence them to hang but to remain in prison for the remainder of their lives. The trial of the century was not to be a trial.[15,16]

Clarence Darrow at the trial.

State's Attorney Robert Crowe prosecuted the case. Crowe had no sympathy for these children of privilege. In his closing argument, he described Leopold and Loeb as "a couple of rattlesnakes, flushed with venom, coiled and ready to strike."

On Monday, July 21, 1924, two months to the date of the murder, the hearing began with State's Attorney Robert Crowe choosing himself to prosecute the case before Judge John R. Caverly, Chief Justice of the Criminal Court, who had served as a judge for 14 years. Darrow saw that he first must convince the judge to let him present witnesses who would tell the court how incidents in their backgrounds made these two highly intelligent boys commit a crime of unthinkable depravity. District Attorney Crowe, in his countering opening statement, argued that alleged mitigating circumstances should instead be listened to by a jury, not a judge. But after Judge Caverly countered that only he need know the extent to which abnormal thoughts dominated the minds of Leopold and Loeb, Darrow would bring before the court several of the United States' most experienced psychiatrists, or "alienists," of the day, including Dr. William Alanson White of St. Elizabeth's Mental Hospital in Washington, D.C.; Dr. William Healy, long associated with the Juvenile Court of Boston; and prominent psychiatrist Dr. Bernard S. Glueck, Sr. Later

*Judge Caverly with Leopold (*center*) and Loeb (*right*).*

he had the court listen to the Chicago practicing physician Dr. Harold S. Hulbert, who with Dr. Karl M. Bowman of Boston on behalf of the boys' parents in mid-June, extensively interviewed Leopold and Loeb. At the heart of Hulbert's testimony was his assessment that both Loeb and Leopold almost totally lacked empathy toward other human beings. Never before had such extensive mental evaluation been introduced into a murder case.[17,18] Finding credible witnesses who would say that Leopold and Loeb also had redeeming aspects of their characters was much harder for Darrow to pull off. Being identified as one of their close friends could mean no future in Chicago for those so identifying themselves. Dad, when asked to testify about Leopold, saw no choice but to turn down Nathan's father's wish for him to testify on Nathan as an ornithologist. His job at LaSalle Extension University would be in jeopardy once he became widely known for being long associated with a perpetrator of what one newspaper described as the "most cold-blooded, cruel, cowardly, dastardly murder ever tried in the history of any court."[19]

[17] Darrow leaned heavily on the psychiatric findings in his closing argument. "The mind, of course, is an illusive thing. Whether it exists or not no one can tell. It cannot be found as you find the brain. Its relation to the brain and the nervous system is uncertain. But when we do find from human conduct that we believe there is a diseased mind, we naturally speculate on how it came about. And we wish to find always, if possible, the reason why it is so. We may find it, we may not find it; because the unknown is infinitely wider and larger than the known, both as to the human mind and as to almost everything else in the universe."

[18] Crowe also hired several psychiatrists to testify, including Hugh Patrick, William Krohn, Harold Singer, and Archibald Church. Their largely anti-Freudian assessments contradicted Darrow's experts. In the end, Judge Caverly discounted all of the psychiatric testimony in making his decision.

After the hearing's end, a *Chicago Tribune* editorial opined that, "Darrow made it known that in his eyes capital punishment is not a proper social response. But they [the editorial writers] were not sure that human progress had reached the stage where a life is not taken for a life." On August 29, the hearing in mitigation or aggravation of the guilty pleas of Nathan Leopold, Jr. and Richard Loeb for the murder of Bobby Franks came to an end. In his closing argument, State's Attorney Robert Crowe said:

> They are no good to themselves. The only purpose that they use themselves for is to debase themselves. They are a disgrace to their honored families and they are a menace to this community. The only useful thing that remains for them now in life is to go out of life and go out of it as quickly as possible under the law.

Darrow's closing argument, which took two days to deliver, became famous and was published in a pamphlet. In his rambling argument, he countered Crowe:

Leopold (center, left) *and Loeb* (center, right) *with their wealthy fathers.*

I have heard precedents quoted which would be a disgrace to a savage race. I have seen a court urged almost to the point of threats to hang two boys, in the face of science, in the face of philosophy, in the face of humanity, in the face of experience, in the face of all the better and more humane thought of age.[20]

Judge Caverly announced his verdict on Wednesday morning, September 10, waiting until halfway through his statement that for the court it would have been the path of least resistance to impose the harshest penalty of the law. He then went on to say "that the court believes that it is within his province to decline to impose the sentence of death on persons who are not of full age."

This determination appears to be in accordance with the progress of criminal law all over the world and with the dictates of enlightened humanity. More than that, it seems to be in accordance with the precedents hitherto observed in this state. The records of Illinois show only two cases of minors who were put to death by legal process—to which number the court does not feel inclined to make an addition.

In conclusion he commented,

Life imprisonment may not, at the moment, strike the public imagination as forcibly as would death by hanging; but to the offenders, particularly of the type they are, the prolonged suffering of years of confinement may well be the severer form of retribution and expiation.

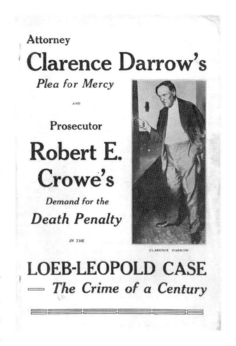

Attorney
Clarence Darrow's
Plea for Mercy
AND
Prosecutor
Robert E. Crowe's
Demand for the
Death Penalty
IN THE

CLARENCE DARROW

LOEB-LEOPOLD CASE
— *The Crime of a Century*

Cover of the widely circulated pamphlet of Darrow's closing argument delivered in the very hot Chicago courtroom. Darrow ended with what was his greatest hope, "that I have done something to help human understanding, to temper justice with mercy, to overcome hate with love," and then topped that off with a quote from the poet Omar Khayyám.

[20] Darrow addressed the death penalty sarcastically in his closing argument. "I can think, and only think, Your Honor, of taking two boys, one eighteen and the other nineteen, irresponsible, weak, diseased, penning them in a cell, checking off the days and the hours and the minutes, until they will be taken out and hanged. Wouldn't it be a glorious day for Chicago?" But he also cannily introduced another argument—"I want to discuss another thing which this court must consider and which to my mind is absolutely conclusive in this case. That is, the age of these boys."

In the end it was the age of the defendants, not the mitigating evidence from Darrow's witnesses, that likely saved the two boys from their widely expected hanging.[21,22]

The next day, on September 11, Leopold and Loeb were transferred under great security to the Illinois State Penitentiary at Joliet. Soon after, Nathan had contacted Dad to ask if they could stay in contact about birds through letter writing. In this way he could maintain his long involvement with the ornithology of the Chicago region. But Dad, now 27, sensed the imperative to remove all aspects of Nathan Leopold from his life as soon as possible and so declined. After the murder Dad quit birding and didn't take it up again for almost 15 years, wanting to do it again when he saw my interest in bird migration; so in the spring of 1938 we started going out to Jackson Park and Eggers Woods.

Then Dad told me that Dickie Loeb was no longer alive, having been killed by a razor deployed by a fellow prisoner in the Stateville Prison several years before. Until then Leopold and Loeb, through acting as model prisoners, had been allowed to use their own monies to supply

[22] The boys were sentenced to 99 years in prison for the kidnapping and to life for the actual murder of Bobby Franks.

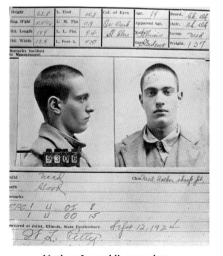

Nathan Leopold's mug shots.

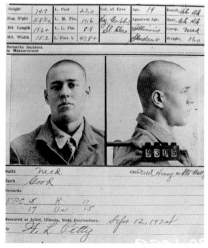

Richard Loeb's mug shots.

themselves and close cellmates with tobacco, candy, and food. The man who killed Loeb was one of them. Leopold, despite for so long being a truly model prisoner, would likely have remained in prison for the remainder of his life if it had not been for his participation in a near end-of-war trial of an anti-malaria drug using prison volunteers. Those taking the new drug were to be later favored when they applied for parole. Leopold, not surprisingly, was virtually routinely turned down in his initial 1953 attempts to be paroled. To aid his cause in 1957, Leopold's autobiography, *Life Plus 99 Years*, was published. In it he made himself into a model prisoner who would pose no potential harm to society were he released.[23,24]

Only five years later, when the famed poet Carl Sandburg came to the hearing to voice support for Leopold, was he granted parole in March 1958 under the condition that for many years he live in a rural region of Puerto Rico that never before had heard of his "crime of the century."

[23] Loeb had been registrar and director of a prison correspondence school. On hearing of Loeb's death, Darrow commented that, "He is better off than Leopold—better off dead."

[24] Leopold retained well-known attorney Elmer Gertz for the 1958 parole hearing; Gertz was also attorney for Jack Ruby and for Henry Miller's *Tropic of Cancer* obscenity trial.

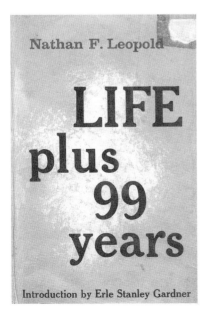

This self-serving book, written in prison, was part of Leopold's campaign for parole, saying, "Thousands of prisoners, especially long term prisoners, look to me to vindicate the rehabilitation theory of imprisonment. I will do my best not to fail in that trust." In this book, Leopold unsurprisingly lays much of the blame for the Franks murder at the feet of Loeb.

BULLETIN 168   AUGUST 1963

CHECKLIST OF BIRDS OF PUERTO RICO AND THE VIRGIN ISLANDS

N. F. LEOPOLD

University of Puerto Rico
AGRICULTURAL EXPERIMENT STATION
Río Piedras, Puerto Rico

Leopold's interest in ornithology never waned, and while in prison he corresponded with at least one of his old bird-watching partners, George Porter Lewis. Once in Puerto Rico, he devoted himself to cataloging the local birds.

Public fascination with the Leopold and Loeb "thrill-kill" case resulted in several Hollywood movies based on the case. Perhaps the most literal retelling of the story was Compulsion, *which starred Orson Welles as the Clarence Darrow character and Bradford Dillman and Dean Stockwell as the boys. Other movies inspired by the case were Alfred Hitchcock's* Rope *(1948) (inspired by Patrick Hamilton's 1929 play* Rope*), Tom Kalin's* Swoon *(1992), and the Barbet Schroeder–directed Sandra Bullock vehicle* Murder by Numbers *(2002).*

The Church of the Brethren, a small Protestant group from Illinois, had built a mission hospital in the village of Castañer, at an altitude of 6,000 feet, about 55 miles southwest of San Juan. There Nathan was to be a medical X-ray technician. Soon he met Trudi Feldman Garcia de Quevedo, the widow of a Spanish physician and owner of a flower shop in San Juan, marrying her in October 1961 after obtaining the permission of his parole board. When in 1963 Nathan won release from parole supervision, they began traveling when Nathan was not working on his *Checklist of the Birds of Puerto Rico*, which was published just after his death from a diabetes-promoted heart attack on August 29, 1971.[25]

[25] On his release, Leopold told reporters that "the only piece of news about me is that I have ceased to be news." He spent the first night of his freedom in the plush 15th-floor "Gold Coast" apartment of his college friend Abel M. Brown with its spectacular views of the Loop and Lake Michigan. Leopold had to cancel plans to fly immediately to Puerto Rico because he suffered from diabetes-induced bouts of nausea during his first days of freedom.

CHAPTER 4

Roosevelt Democrats (1925–1942)

Dad married my mother, Margaret Jean Mitchell, on May 23, 1925. While to her mother's family she was long called Bonnie Jean, in high school and later at work she began calling herself Jean, never liking the name Margaret given to her at birth. In January 1919 she graduated from Wendell Philips High School at 39th and Prairie Avenue. Then she went to the University of Chicago (U of C) for two years until there was no more family tuition money. Though her Glasgow-born father Lauchlin (also spelled *Laughlin*) Alexander Mitchell had once been one of Chicago's more prosperous merchant tailors, the arrival of ready-made suits eroded his business, and by the time of his untimely death from a horse-related accident in 1914 outside the Palmer House Hotel, there was little money to pass on toward mother's college education.[1]

Already in high school, my mother's physical activities were compromised as a consequence of a year-long, late-grammar-school bout of streptococcal rheumatic fever. The resulting leaky heart valves permanently made her short of breath upon even the mild exertion of going up stairs. Only when my mother later died from heart failure did I learn that she was in fact born on Christmas Eve in 1899, not in 1900. Not wanting to explain to potential employers that a damaged heart had lengthened her passage through school, she took a year off her age upon joining the working force. This fib, however, was trivial compared to that perpetuated by her Indiana-born mother, Elizabeth (Lizzie) Gleason, whose parents, born Michael Gleason and Mary Curtin, had emigrated from Tipperary, Ireland, during the potato famine of the late 1840s. She

Bonnie Jean Mitchell.

[1] Wendell Phillips Academy High School, which opened in 1904, was named for the abolitionist Wendell Phillips and was an early racially mixed high school in Chicago. By 1930, it was predominantly African American.

Lauchlin Mitchell's death certificate. He had been hospitalized for 33 days and finally succumbed to pancreatitis at the age of 59, according to this document.

Lauchlin Alexander Mitchell (1855–1914), son of Robert Mitchell and Flora Mac-Kinnon. Painted at the time of the World's Columbian Exposition.

long successfully hid from her merchant-tailor husband that she was, in fact, born on the family farm in the Michigan City outskirts in 1861, not, as he was long led to believe, in 1875. Though Lauchlin Mitchell long supposed he was 20 years senior to his wife, he was, in fact, only six years older. Behind my grandmother's big lie was the need to hide the fact that for 15 years she had been kept in high style in Chicago by the wealthiest man in Michigan City, John H. Barker, whose large factory on its west side had the potential to assemble 15,000 freight cars a year. Upon leaving her farm home, she had first entered his 631 Washington Street home in Chicago as a live-in domestic, later becoming his wife Genia Brooks's traveling companion during her stays in more southerly spas. That my mother was strikingly beautiful likely reflected the once equally good looks of her mother.[2]

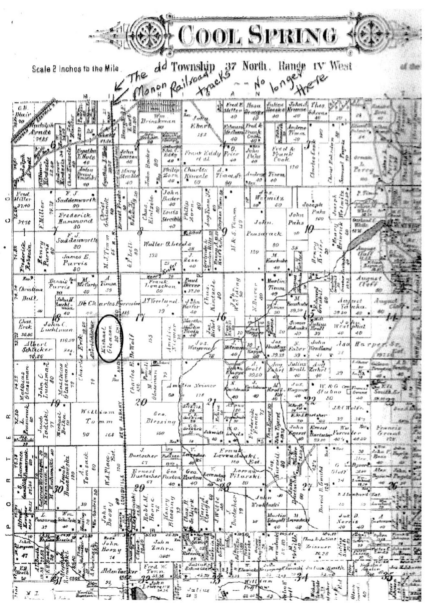

2 The very wealthy John H. Barker (1844–1910) started as a grocery clerk in Chicago but moved back to Michigan City, Indiana, eventually becoming president of Haskell & Barker Car Company, which later merged with the Pullman Company. He was also director of several banks and was the mayor of Michigan City from 1880 to 1882. His estate was estimated at $50 million to $60 million at his death.

Plat of Cool Spring Township in LaPorte County, Indiana, showing Michael Gleason's property bordering the Louisville, New Albany, and Chicago Railroad tracks. Just east of Gary, Indiana, Cool Spring was very close to the shore of Lake Michigan.

Bonnie Jean Mitchell.

"Lizzie" Mitchell in 1925. She joined the Watson household in 1933.

Mother and Dad met each other at work with mother's first job after her U of C years being in the personnel office at the LaSalle Extension University, whose building at Michigan Avenue at 41st was located exactly across the street from where she grew up in a three-story brick apartment. Later mother went to work via a 10-minute streetcar trip after her mother moved to a less expensive multibedroom apartment at 59th and Indiana just across Washington Park from the U of C. To keep afloat financially, Lizzie took in paying guests, effectively running an Irish boardinghouse.[3]

[3] LaSalle Extension University, founded in 1908, offered courses in accounting and law. Federal Trade Commission lawsuits forced the eventual closure of LaSalle in 1980. An early pioneer for distance learning, LaSalle provided the opportunity for higher education for those who could not afford a college education. Adlai E. Stevenson, grandfather of the presidential candidate and himself Vice President of the United States (1893–1897), helped establish LaSalle. In 1923 the school claimed an annual enrollment of 60,000 and had 1,500 employees.

LA SALLE FOLKS

| Vol. VIII | Chicago, March 17, 1923 | No. 11 |

"SMILING VOICE" WHO WELCOMES NEW GIRLS
The Employment Department and the Lady at the Wheel

A PLEASANT ripple in her voice, a friendly tilt to her head—instinctively you like her. This is the lady of this week's center picture. Somehow, if you are the least bit inclined to be "gabby" (you know), and most of us are somewhat, you find yourself telling her about the time Johnny fell downstairs while chasing the cat, what sort of a boss you once had at the Catchem & Killem Collection Agency, and how you happened to start the mad career of stenography.

No, her hair isn't smoothed down just for the occasion. Her hair is always that way. We have offered Bloodhound Bill, the special detective, a bonus if he will find her hair untidy. He has never won that bonus. He reports that two ruffled things about Jean Mitchell "am not"—her hair and her disposition. We always think what a satisfactory little girl she must have been to her mother, for she always must have been one of those tidy little angels that never pulled poor kitty's tail.

That about kitty gives us another thought. All kittens are especially fond of Miss Mitchell, possibly reflecting her feeling toward them. The LaSalle Cat has adopted her as Godmother and Godfather for her family, and decided to share Room 7 with her. You know what they say about animals and children—and Mr. Deniston tells us that sign never fails.

Many girls, when looking for just the right position in business, dread the ordeal of talking to some Ogre Employment Man. They sit on the bench outside and quake and tremble, and then when the time comes for the interview they open their mouths like fishes, and therefrom no sound issues forth. We've all been there at one time or another, but never in LaSalle.

LaSalle's X-Ray

FOUR thousand letters and over come to LaSalle's Mailing Department daily. These letters are opened, taken from their envelopes, and routed to the proper department. Several times a letter received by some department reads: "Enclosed, find check in payment of my account"—but no check was attached.

This would have caused considerable expense and trouble but for an invention of our efficient Mailing Department head of the "Letter-X-Ray"—a glass-top box. In the box is placed an electric light, which shines thru the glass top. After the letters have been removed from their envelopes, the envelopes are passed over this glass top of the "Letter-X-Ray." The light shines thru the envelopes, and, by making them translucent, shows objects within.

R. A. McLennan

Here, girls may do the "before-quaking" act, but once they sit down beside Miss Mitchell's tidy desk, and receive the full sunshine of her 5,000 volt smile, it's all off. She's got their hearts for ever 'n ever. They begin to blossom out and tell all about themselves, their pet hopes, ambitions, and abilities; also (incidently, and perhaps subconsciously, as our psychologists say) about their shortcomings and faults. Because of this tact, which puts girls at their ease, Miss Mitchell is able in a short time to decide which girl will be the lucky one to be admitted to the little family of LaSalle.

You men know that the new girls are "nice" and fit into the work. Girls know that "there is a dandy bunch of new girls in the department," but dear people, the credit goes to Our Lady of the First Interview. To her, and her charm, can be given the credit of gathering a group of efficient and suitable girls and women into the LaSalle family. She is responsible for that all-important first impression, which inspires our girls to deliver the best they have to LaSalle. In the end, the big, important end, the results go to our students. Each "weetsy, beetsy" service, each "giant, whopper" service, that is delivered to any one of our students reflect Miss Mitchell's voice and magic touch.

Editorial Note: Here we will change the record from minuette to stately march, and hear what Mr. Deniston has to say about Miss Mitchell and her job.

Let us step into Room No. 7. Here sits Miss Jean Mitchell, who functions as the medium between the various departments of the University and the many sources of help supply, in selecting competent, alert, responsible and ambitious women employes.

When there are positions open for young women, she must see that they are thoroly and properly filled. To do this, it is necessary for Miss Mitchell to have an intimate knowledge of the work in each department and in each position. She must be familiar with the particular requirements of each position open, to select persons of special qualifications to fill the positions, and to give our student members the 100 per cent service they must have.

She must be fully familiar with the various sources, from which to draw likely applicants to fill the vacancies as they present themselves. Some of the sources are present employes and their friends, previous employes, voluntary applications, help-wanted advertisements, schools, and private and public agencies.

Before drawing upon outside sources for desirable applicants, Miss Mitchell carefully considers the promotion of present employes to positions suited to them. If a transfer is not advisable, applicants must then be brought in from the outside.

The next step is to determine the suitability of new applicants for employment. This is done by interview, observation and tests to decide on their mental, physical, and temperamental qualifications for the positions offered. The desirable applicants are then referred to department managers.

Not only does Miss Mitchell secure, interview, and recommend applicants for employment, but she supervises the building up and maintenance of a complete history of past, present, and potential LaSalle employes. She closely studies employment conditions thruout the country for the purpose of determining the effect of such conditions on our work at LaSalle.

This young woman, upon whose daily judgment much depends, is well prepared for this work, by training at the University of Chicago, and by actual experience here at LaSalle since October 1920. She has for some time interested herself in settlement work and the helping of others, and thru her pleasing personality has made hundreds of staunch and loyal friends here at LaSalle.

LaSalle Folks article describes Jean Mitchell's work at LaSalle and notes that she "thru her pleasing personality has made hundreds of staunch and loyal friends here at LaSalle."

1915 ad for LaSalle Extension University. The school was modeled on the U of C's home study course, an idea of U of C president William Rainey Harper. This populist-aimed ad exhorted potential students to enroll—"The working man needs you to fight his legal, social, and economic battles."

Upon their marriage, Dad and Mother moved into an apartment building in Hyde Park near where his parents and two brothers Bill and Stanley lived. For their honeymoon they went to Niagara Falls and a year later to Quebec, where they stayed in the huge Château Frontenac in Quebec City. Their last lengthy travel before I was born on April 6, 1928, was to the Catskill Mountain–sited, stone-clad, eight-story Mohonk Mountain House, which provided guests the opportunity of

Newspaper and formal wedding announcements of marriage of Bonnie Jean and James.

reading from the many thousands of books on shelves surrounding the dining room. After my sister Elizabeth Laughlin (Betty) was born on June 23, 1930, my parents moved five miles south to an apartment on Merrill Avenue off 79th Street. It was big enough to also accommodate my cash-strapped grandmother, whom Betty and I would later always call Nana. By then her previously high-yielding Argentine bonds were effectively worthless, taking away any possibility of living alone.

For their early years of marriage, Dad and Mother could avail themselves of a large black Hudson roadster that let mother in 1932–1933 bring me to the prekindergarten division of the U of C's John Dewey–founded University High School as well as let Dad regularly play golf in Jackson Park. Dad's love of golf went back to his high school days in La Grange, where his mother had purposely chosen to live next to its new golf club. Then Dad's dress still reflected his early semiprivileged years,

Lake Mohonk Mountain House.

Bonnie Jean in mid-1920s.

James D. Watson, Sr., in mid-1920s.

Jean Watson and her infant son James, born in 1928.

James and his young son James, Jr.

State _Illinois_ County _Cook_ Ward of city _Seven_ Block No. _25_ Incorporated place _Chicago city_

DEPARTMENT OF COMMERCE—BUREAU OF THE CENSUS
FIFTEENTH CENSUS OF THE UNITED STATES: 1930
POPULATION SCHEDULE

Enumeration District No. _16-304_ Supervisor's District No. _29_ Sheet No. _7-A_ 141

Enumerated by me on _April 8_, 1930. _John J. Wildeson_, Enumerator.

1930 U.S. Census showing Watson household.

wearing knickerbockers (knickers) when touring or playing golf while mother wore long skirts covering almost well below the knee.

My parents' comfortable way of life, however, came to an abrupt halt when in 1933 the Depression slashed Dad's salary by 50% to $3,000, forcing the sale of their large Hudson roadster. I was last a passenger in it in early 1933, when even the big banks began to fail and Mother drove to the still-mighty LaSalle National Bank in the Loop to take out the small sum of family money she had deposited there. From then on Dad commuted to work by streetcar, sometimes changing at 63rd and Stoney Island to more quickly reach by the "L" (elevated) his Michigan Avenue place of work. In the summer of 1933, the five of us moved two blocks to the east into a modest, one-storied 7922 Luella Avenue bungalow containing two bedrooms, the smaller one next to the kitchen being occupied by Nana, who increasingly did much of the cooking to ease the strain on my mother's heart. To create a new room in the attic for my sister and me, insulation was placed between the roof's rafters. As we grew into adolescence, a second room was built using the window facing our back garden bounded by a garage on the long alley running between Luella Avenue and Paxton Avenue to our west.

Dad's two-year-younger brother Bill lived then in New Haven, Connecticut, to which he, his wife Betty, and their daughter Ruth moved upon his becoming an Assistant Professor of Physics at Yale. In 1936, they returned to Chicago for the funeral of my paternal grandmother,

Watson family Hudson roadster.

Jim (right) and sister Betty on the front porch of their 7922 Luella Avenue home.

Nellie Dewey Ford Watson, who died soon after being diagnosed with ovarian cancer. By then the only son still living with her in her Dorchester Avenue flat was my father's youngest brother, Stanley, who went directly from high school to work in business, eventually becoming a manager of a community bank at the intersection of Jeffery Avenue and 71st Street through whose center ran the South Shore Branch of the Illinois Central (IC). When Mother saw a need for Betty and me to be well dressed, she took us to the Loop to buy clothes at the upscale Carson Pirie Scott department store. We used the IC even though its tickets cost almost twice that needed to get to the Loop by the much more time-consuming streetcar.

Living nearby in South Shore was Dad's brother Tom, whose obvious like of good food led to his being a chef in a succession of Chicago-area restaurants that only briefly survived. More than six feet tall like all his brothers except for my 5'10" father, Tom weighed over 250 pounds even in his thirties, possibly inheriting his tendency for obesity from his mother Nellie, whose weight rose to nearly 300 pounds in her later years. While Tom and his wife, Etta's, two sons had normal physiques, their sisters, Virginia and Nan, had the bad luck of growing up obese. Tom annually hosted a sumptuous Thanksgiving turkey meal for the South Shore Watsons—his family and ours and his brother Stanley, who remained single until his early thirties. Only in 1939 did he marry the divorcée Anna Mary Schoof, whose two-year-old son Charles also took on the Watson surname.

Exactly how long back into his past Dad broke with his Republican Party–orientated father and brothers in becoming a Democrat I do not remember. When Franklin Roosevelt became President on March 4th, 1933, Dad was already an impassioned exponent of social activism, believing that without governmental intervention those without jobs and their families had very grim fates. Out-of-date, early-1930s copies of the

New Masses were long piled up in our basement. By the time I began to read newspapers, Dad's political interests were much better served by weekly readings of *The New Republic* and *The Nation*.[4]

My first political exposure came from not wearing Landon for President sunflower buttons to grammar school in the run-up to the 1936 presidential election. The vast majority of Horace Mann students had Republican parents, the Democratic component of our neighborhood largely being families of Irish heritage, whose children went to Our Lady of Peace parochial school. Joining me in not wearing sunflowers were the twin Press girls, Norma and Violet, and my closest buddy, Charles Frankel, who went on to be a newspaper reporter in Honolulu. They all came from liberal Jewish families. Mother's Democratic credentials later got her selected as a non-voting member of the Indiana delegation to the 1940 Democratic National Convention at the Chicago Stadium. There she rooted for the handsome Indiana Governor Paul V. McNutt to become FDR's vice-presidential running mate. But FDR wanted and got his Secretary of Agriculture, the Iowa corn breeder Henry A. Wallace, chosen to run with him.[5]

Ticket to the 1940 Democratic National Convention.

All through the time we were living on Luella Avenue, Mother remained in close contact with her mother's family, all of whom still lived in Michigan City, Indiana. Once we were passengers in an early 1900s steamship that still took summer vacationers from the dock at the mouth of the Chicago River to vacation sites along the eastern shore of Lake Michigan. Most often we were passengers on the South Shore and South Bend Railroad electric trains that passed through Hammond and

[4] The *New Masses* was started in 1926 by the Workers Party of America and was considered the voice of the American political left. Many prominent writers, including Theodore Dreiser, John Dos Passos, Dorothy Parker, Ernest Hemingway, and Eugene O'Neill, wrote for the *New Masses*. Its increasingly Marxist stance and the post–World War II political climate led to its eventual demise in 1948.

[5] The 1940 Convention marked FDR's bid for an unprecedented third term as President, and he was nominated on the first ballot. Unfortunately, McNutt only received 68 votes to Wallace's 626 votes.

Loretta Olvaney, Geraldine Olvaney, and Margaret Jean Mitchell on the steamer S.S. Theodore Roosevelt. The ship, which was built in 1906, plied the waters of Lake Michigan. It was taken over by the U.S. Navy in 1918 to serve as a troop transport in World War I and eventually came back to Lake Michigan in 1927.

Gerald Olvaney's parents John and Catherine.

Gary while making the 60-minute journey to Michigan City. Mother was closest to her first cousin, Gerald Olvaney, a son of Nana's sister Catherine. Through owning a successful grocery store, Gerald had a car that regularly took us fishing to several of the small lakes that were dotted all over northern Indiana. Almost always enough perch and bluegills were caught to provide for the evening meal. Long wanting to be a radio actress was his oldest daughter Geraldine, whom Mother especially liked because she remained a Democrat when the rest of her family had moved to the Republican Party. To audition for possible acting roles, Geraldine occasionally came to Chicago's Loop, afterward coming out to our house. Unfortunately she never moved from being an amateur to a paid professional and though inherently good looking, but maybe a bit too intelligent, never found a man to marry.

Mother's social network steadily expanded through her becoming the Democratic captain of our precinct in the 7th ward. In contrast, Dad's social life was always more limited. His key friend was Linton Keith, whose faraway southwest side Palos Park home sadly restricted the occasions they could get together to talk about ideas. Their friendship went back to the first years of Dad and Mother's marriage, when they and Linton and his wife, Jo, spent long evenings together discussing books with philosophical overtones as to how the good life should be reached. Though early in life Linton accepted religious dictates, by then he prided himself as a Humanist whose rules for daily life emerged from human nature and dignity.

Dad likewise saw the need for philosophically driven principles to replace the religious revelations that governed most American lives. Then he was increasingly influenced by the philosopher and educator John Dewey (1859–1952), with whom he likely shared an early 18th century New England Dewey ancestor. Early in his career Dewey had taught at the U of C before moving in 1904 to Columbia University for the remainder of his career. Like the American psychologist William James, Dewey was an articulate proponent of Pragmatism, in which the practical consequence of action is primary. Whenever Dad read discussions of his ideas in newspapers or magazines, he invariably clipped them out, leading to a Dewey file folder even fatter than his Walt Whitman, Joseph Conrad, or John Cowper Powys files.[6,7] Dad also much enjoyed interacting socially with our neighbor across the street, the Belgian-born Jew, Rene Bender, whose job with the Sinclair Oil Company would later send him to Ethiopia. On many days Dad had to bring accounts home to let him finish his day's work after supper or on weekends. The good side of his almost hour-long commute to and from work was the time it gave

[6] "To me an interesting person is one alive with intellectual curiosity." —JDW, Sr., March 4, 1939, diary entry

[7] "I sometimes think there is no new experience in life. Something may happen to you that you think has never happened before, but you are mistaken. You have only to see or hear or smell or feel something and you will discover this experience that you thought was new has happened before." —JDW, Sr., March 4, 1939, diary entry

[8] "In general it can be said, I think, that he who reads is a safe man to tie to. Certainly he makes a delightful companion. After all, its what we retain from our reading that helps to make us interesting and able to understand." —JDW, Sr., April 11, 1939, diary entry

[9] "There is a kind of optimist who maintains that one should never be bored, that it indicates a defect on the part of the bored person. One ought, the optimist says cheerily, to find interest in anything and everything. This is nonsense. People ought to get bored much quicker than they do, and insist on having something interesting. I would say that the civilized man is distinguished by his capacity for being bored and finding a remedy for it in thought and action." —JDW, Sr., March 16, 1939, diary entry

him for the reading of books, magazines, focused literary topics, natural history, and politics.[8,9]

Dad's bird-watching passion only seriously resumed when I reached 10 and had legs sufficiently long to keep up with his pace. Lack of a car then largely confined us to streetcar trips to Jackson Park and to the Wolf Lake Preserve (Eggers Woods), where much earlier he had gone as a passenger in Nathan Leopold's red roadster. In its swamps we had excellent chances of seeing, say, the Woodcock or Wilson's Snipe during spring vacations. Insect-eating Warblers and Vireos were best spotted in Jackson Park, whose trees, all planted after the conclusion of the 1893 World's Columbian Exposition, had not reached heights beyond the range of Dad's bird glasses. Lack of money kept him from ever moving up from the old-fashioned rigid binoculars of his youth. This deficiency never kept Dad from being right on his bird identifications most often done from their vocal outpourings.[10]

Eggers Woods bird list by budding ornithologist James D. Watson, Jr.

LIST OF BIRDS SEEN BY J.D.WATSON AND J.D.WATSON,JUN., in 1938

No.	Bird	Date	Locality	Com. or R
1	English Sparrow	Jan. 1st	Everywhere	Abundant
2	Starling	" "	Jackson Park	2
3	Mallard Duck	" "	Jackson Park	Common
4	Wood Duck	" "	" "	3
5	Scaup Duck	" "	" "	1
6	Pintail Duck	" "	" "	1
7	Redhead Duck	" "	" "	3
8	Snow Geese	" "	" "	3
9	Canada Geese	" "	" "	5
10	Golden-eye Duck	" "	" "	3
11	Herring Gull	" "	" "	12
12	Downy Woodpecker	Jan. 2nd	Eggers Woods	3
13	Tree Sparrow	" "	" "	Common
14	Junco	" "	" "	6
15	Cardinal	" "	" "	1
16	Blue Jay	" "	" "	1
17	Redpoll	" "	" "	2
18	Chickadee	" "	" "	1
19	Goldfinch	" "	" "	5
20	Hermit Thrush	Jan. 16th	" "	1
21	Pine Siskin	" "	" "	1
22	Blue Geese	Feb. 20th	Jackson Park	5
23	Green-winged Teal	" "	" "	1
24	Buffle-head Duck	" "	" "	1
25	Prairie Horned-lark	Feb. 22nd	Eggers Woods	25
26	Marsh Hawk	" "	" "	1
27	Robin	Mch. 12th	Jackson Park	12
28	Bronzed Grackle	" "	" "	5
29	Crow	" "	" "	45
30	Killdeer	" "	" "	2
31	Bluebird	" "	" "	2
32	Golden crowned Kinglet	Mch. 19th	" "	3
33	Red-winged Blackbird	" "	" "	10
34	Flicker	" "	" "	1
35	Fox Sparrow	" "	" "	2
36	Vesper Sparrow	Mch. 20th	Home	2
37	Meadow Lark	Mch. 20th	Eggers Woods	15
38	Song Sparrow	" "	" "	7
39	Red-tailed Hawk	" "	" "	1
40	Cowbird	" "	" "	20
41	Phoebe	" "	" "	2
42	Towhee	" "	" "	3
43	Migrant Shrike	" "	" "	1
44	Bonaparte Gull	Mch. 21st	Jackson Park	5
45	Myrtle Warbler	Mch. 22nd	Home	1
46	Pied-billed Grebe	Mch. 26th	Jackson Park	1

[10] "We met Stein in the woods and he pointed out a Rough-legged Hawk…. Not a common hawk by any means…. I pointed out to Jimmy the alarm note of the Song Sparrow. It is a metallic chip—very distinctive when once learned…. A few days ago hardly a bird, not a sound; everything rigid and severe; then in a day or two, the barriers of winter give way, and spring comes like an inundation, and the birds keep pace. These days one cannot help but be happy." —JDW, Sr., March 25, 1939, diary entry

1938 bird list compiled by James Watson, father and son. (Figure continued on following page.)

- 2 -

47	Kingfisher	March 26th	Jackson Park	1
48	Brown Creeper	" "	" "	1
49	Red-breasted Merganser	" "	" "	4
50	Ring-billed Gull	" "	" "	2
51	Sparrow Hawk	March 29th	" "	1
52	Long-eared Owl	" "	" "	1
53	Rusty Blackbird	" "	" "	6
54	Swamp Sparrow	April 9th	Eggers Woods	2
55	American Bittern	" "	" "	1
56	Blue-winged Teal	April 12th	Jackson Park	2
57	Loon	" "	" "	1
58	Black-crowned Night Heron	" "	" "	1
59	Chipping Sparrow	" "	" "	1
60	Sharp-shinned Hawk	" "	" "	1
61	Ruby-crowned Kinglet	" "	" "	3
62	Sapsucker	April 13th	Home	1
63	Brown Thrasher	April 15th	Jackson Park	1
64	American Merganser	" "	" "	1
65	Purple Martin	April 17th	Eggers Woods	4
66	Great Blue Heron	" "	" "	1
67	Coopers Hawk	" "	" "	1
68	Field Sparrow	" "	" "	50
69	White-throated Sparrow	" "	" "	5
70	Winter Wren	" "	" "	1
71	House Wren	" "	" "	1
72	Mourning Dove	" "	" "	2
73	Purple Finch	" "	" "	3
74	Tree Swallow	" "	" "	1
75	Wilson Snipe	" "	" "	1
76	Barn Swallow	April 20th	Jackson Park	1
77	Coot	" "	" "	1
78	Blue-grey Gnatcatcher	" "	" "	1
79	Bank Swallow	April 24th	" "	1
80	Hairy Woodpecker	April 25th	Home	1
81	Solitary Sandpiper	April 27th	Jackson Park	1
82	Spotted Sandpiper	" "	" "	2
83	Little Green Heron	" "	" "	1
84	Palm Warbler	" "	" "	6
85	Yellow Warbler	" "	" "	1
86	Cape May Warbler	" "	" "	1
87	Red-headed Woodpecker	" "	" "	1
88	Chimney Swift	" "	" "	3
89	Grasshopper Sparrow	" "	" "	2
90	Black & White Creeper	" "	" "	1
91	White-crowned Sparrow	" "	Home	2
92	Olive-backed Thrush	April 29th	Home	1
93	Sora Rail	May 1	Eggers Woods	2
94	Long-billed Marsh Wren	" "		10

1938 bird list continued from previous page.

- 3 -

95	Indigo Bunting	May 3rd	Jackson Park	1
96	Baltimore Oriole	" "	" "	3
97	Rose-breasted Grosbeak	" "	" "	2
98	Scarlet Tanager	" "	" "	1
99	Lesser Yellow-legs	" "	" "	1
100	Water thrush	" "	" "	2
101	Red-breasted Nuthatch	" "	" "	3
102	Parula Warbler	" "	" "	1
103	Tennessee Warbler	" "	" "	1
104	Chestnut-sided Warbler	" "	" "	1
105	Black-throated Green "	" "	" "	1
106	Crested Flycatcher	" "	" "	2
107	Warbling Vireo	" "	" "	1
108	Kingbird	" "	" "	1
109	Redstart	May 6th	" "	1
110	Bobolink	May 8th	" "	2
111	Catbird	May 9th	" "	1
112	Ruby th. Humming-bird	" "	" "	1
113	Magnolia Warbler	May 10th	Home	1
114	Gray-ch. Thrush	" "	"	1
115	Nashville Warbler	May 11th	Jackson Park	2
116	Pigeon Hawk	" "	" "	1
117	Northern Yellow-throat	" "	" "	1
118	Wilson Warbler	" "	" "	1
119	Ovenbird	May 13th	" "	1
120	Woodthrush	" "	" "	1
121	Orchard Oriole	" "	" "	1
122	Least Flycatcher	" "	" "	1
123	Black Tern	May 15th	Eggers Woods	10
124	Black-billed Cuckoo	" "	" "	1
125	Blackpoll Warbler	" "	" "	4
126	Short-billed Marsh Wren	" "	" "	1
127	Red-eyed Vireo	" "	" "	1
128	Canadian Warbler	May 18th	Jackson Park	1
129	Common Tern	" "	" "	4
130	Least Bittern	" "	" "	1
131	Wood Pewee	" "	" "	1
132	Bay-breasted Warbler	" "	" "	1
133	Blackburnian Warbler	" "	" "	1
134	Prothonatary Warbler	" "	" "	1
135	Trails Flycatcher	" "	" "	1
136	Yellow-billed Cuckoo	" "	" "	1
137	Nighthawk	" "	Home	3
138	Red-sh. Hawk	May 20th	Jackson Park	1
139	Black-Th. Blue Warbler	May 21st	Home	1
140	Kentucky Warbler	" "	"	1
141	Sanderling	May 25th	Jackson Park	1
142	Cedar Waxwing	" "	" "	10
143	Yellow-bellied Flycatcher	" "	" "	1

1938 bird list continued from previous page.

- 4 -

144	Cliff Swallow	May 30th	Sag Canal	3
145	Savanna Sparrow	June 12th	Cherry Hill Crs.	2
146	King Rail	June 20th	Eggers Woods	1
147	White-br. Nuthatch	June 22nd	Palos	1
148	Tufted Titmouse	" "	"	1
149	Dickcissel	" "	"	2
150	Semipalmated Plover	Aug. 29th	Jackson Park	2
151	Semipalmated Sandpiper	" "	" "	5
152	Caspian Tern	" "	" "	1
153	Yellow-throated Vireo	Sept. 4th	Eggers Woods	1
154	Woodcock	Sept. 24th	" "	1
155	Wormeating Warbler	Sept. 24th	" "	1
156	Turkey Vulture	Oct. 2nd	Turkey Run, Ind.	30
157	Carolina Wren	" "	" "	1
158	Pectoral Sandpiper	Oct. 15th	Home	10
159	Broad-winged Hawk		Eggers Woods	1
160	Hooded Merganser	Dec. 2nd	Jackson Park	1
161	Ruddy Duck	Dec. 9th	Jackson Park	1

Jim Watson at age 11 with his father and his sister, Betty.

Our grandest birding experience together came in the summer of 1939, when an acquaintance of Uncle Stanley drove our family to California and back. Dad paid for the fuel and related expenses while the driver in return saw his girlfriend during the five days we were in Los Angeles. Even before we got to Rocky Mountain Park outside Denver, I delighted in my first sightings of Western Meadowlark, Lark Bun-

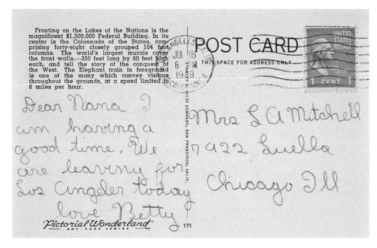

Postcard sent by Betty from the California World's Fair in San Francisco in July of 1939 to her Nana Elizabeth Mitchell, at home in Chicago.

tings, and Yellow-Headed Blackbirds along the "Lincoln Highway" of Iowa and Nebraska. Later I found the big blue Steller's Jays of the Rockies and Sierras much more splendid than our Eastern Blue Jays. San Francisco let us add Pelicans to our lifetime lists, while the main treat of the drive down to LA was the massive trunks of the giant Sequoia on the western slopes of the Sierras. The low point of the trip was a day-long delay at a gas station in the Mojave Desert needed to remove a supposed gas-saving carburetor installed in LA that was to lower the gasoline consumption. Afterward we excitedly drove over the Boulder Dam (now the Hoover Dam) into Nevada and then on to Zion and Bryce Canyon National Parks of Utah before swiftly driving back to Chicago to let the entire trip be done within Dad's two-week annual vacation period.

All through my later grammar school days, Dad and I threw hard baseballs to each other, he still retaining the mitt from his LaGrange boyhood and I a shiny new one that I got as a cherished April birthday present for immediate spring use. A large vacant corner lot across 79th Street was perfect for softball games, where I could put more spin on softballs than bigger boys who threw faster. Though lack of money kept Dad and me from going more than once a year by the "L" to Cubs games at Wrigley Field, I spent most of my summer afternoons listening on the radio to Cubs games. Dad's lack of any trace of interest in the much-closer-to-home White Sox baseball at Comiskey Park likely most reflected his still strong memories of the two-decade-before, gambling-promoted, Black Sox blowing of the 1919 World Series.[11]

All through the 1936–1939 Spanish Civil War, Dad hoped against hope that the leftist, anticlerical, Loyalist armies (the Spanish Republican Armed Forces) would prevail over General Francisco Franco's right-wing Fascist (Nationalists) forces. Though Dad was increasingly seeing the Russian Communist State as an awful Stalin-led dictatorship, he was even more repelled by Spanish Catholicism, seeing it as a reactionary

[11] "Is not the problem of life much more simple than people think? I think it is. I have gotten along without any formal religion for many years. Life has interested me at all times, and I've never yet had time to do all the things I have wanted to do, nor to read all the books I've wanted to read. This is a very confusing epoch to many, but I am not confused. There may or may not be an intelligence back of the universe. It really makes no difference. I neither look for nor desire a hereafter. For me, living wholly and utterly each day is quite enough." —JDW, Sr., March 28, 1939, diary entry

force invariably in bed with the privileged hereditary wealth.[12] Dad viewed the anti-Semitism of so much of the Christian world as grotesque, finding himself instinctively in awe of the polymath Jewish intellectuals who used the radio and books to discuss the beauty and power of ideas.

Only disgust did he feel for the German Reich's anti-Semitism and for their American equivalents who kept Jews away from Christian summer resorts, one of the most famous being the Grand Hotel on Mackinac Island where Lake Michigan and Lake Huron meet. With time a number of much-more-modest grand hotels sprang up on Lake Michigan's eastern shore to cater to the summer needs of the Chicago-area Jews. Many were in South Haven, Michigan, some 15 miles away from the summer-guest-taking farm to which, in 1940, my parents took Betty and me for a much anticipated, week-long escape from Chicago's summer hot days. During our visit, an extended Jewish family came briefly into our lives when they occupied an adjacent cabin and joined the communal supper held in the farmhouse's dining area. After supper singing began and Mother's good voice let her take the lead to sing a catchy verse about Eastern European immigrants that kept repeating a Finkelstein motif. What Mother saw as harmless ethnic teasing, Dad saw as blatant Chicago Irish anti-Semitism. Later back in our cabin, he openly voiced his strong disapproval of Mother's most inappropriate song selection. It led to the Jewish family's return to South Haven the next morning.

Though Betty and I both were given lessons by local piano teachers, neither of us had any underlying talent. While replacing our old-fashioned upright piano with a baby grand inherited from Dad's mother made our living room look better, it did not make our fingers move effectively. My lessons abruptly stopped when I totally panicked during a recital not remembering how to end my piece. Betty's lessons went on longer but she never seriously improved, say, to being able to sight-read even moderately difficult compositions.

[12] The Spanish Civil War effectively ended with the fall of Madrid on March 28, 1939, after a three-year siege by Franco's Nationalists.

Jim Watson, age 12.

Betty and Jim at the piano.

[13] "Jean and the children came downtown, and after lunch, we traipsed thru the rain to the Art Institute to see the Picassos, become modern, and shake off our old-fashionedness. They have gathered maybe 400 Picassos of all sizes, shapes and periods into the galleries and I have to confess that I looked at many of them with wonder and lack of understanding.... However, to my untutored eye, it was apparent that the pictures were the work of a master. How unusual it is that he did not have to die to have the world recognize him."
—JDW, Sr., March 2, 1940, diary entry

Afterward Mother took us into the Art Institute on Michigan Avenue for drawing lessons from Dad's uncle, the multi-talented Chicago artist and cultural force, Dudley Crafts Watson. His lessons started with our being given charcoal sticks and large pieces of paper onto which we were to draw clothed models who sat on the stage. Sadly Dudley's lessons reaffirmed to Dad and Mother what was already too obvious from our grammar school art classes—that neither of us had much artistic talent. Our visits to the Art Institute, however, let me repeatedly wander through museum rooms, starting to much enjoy its outstanding 19th century French Impressionist paintings, particularly, the huge Seurat *A Sunday on La Grande Jatte* as well as its Chester Dale collection of rose and blue period Picasso figures.[13]

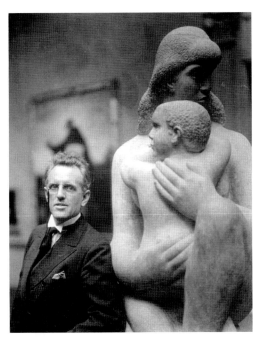

Dudley Crafts Watson at the Chicago Art Institute next to a sculpture by famed artist William Zorach. Dudley moved back to the Art Institute of Chicago in 1924 and remained there for 36 years as official lecturer. This photo was taken by Jessie Tarbox Beals, a pioneering female photojournalist.

[14] Dudley's lectures were very popular with the public—as entertaining as they were informative; in 1943 he gave 370 talks to about 120,000 people at the Art Institute. Dudley also traveled the country lecturing on art, and he led art-oriented tours—"Art Pilgrimages"—to Europe and Latin America for many years. Always interested in how music and painting complemented each other, he presented his lectures "with three dimensional slides in their true colors with correlative music" and was credited as the "creator of the music picture symphony." He continued this theme with his paintings, which he described as "Music for the Eye." He said he "composed paintings the way musicians compose music."

Dudley, at my Mother's urging, drove to South Shore early in 1941 to lecture before a well-attended annual meeting of the Horace Mann Parent-Teacher Association (PTA).[14] Mother was then its president and delighted in Dudley's catching on paper the likeness of virtually anyone presented to him to draw. Afterward Dudley talked about his regular phone conversations with his much younger cousin Orson Welles (born in 1915 in Kenosha, Wisconsin). By then Orson's fame was everywhere, he first bursting into the public imagination through his October 30, 1938, Mercury Theatre radio presentation of H.G. Wells' *War of the*

[15] *The New York Times* provided extensive coverage of Welles's electrifying Mercury Theatre March 31, 1938 *War of the Worlds* broadcast. In a 1955 "Sketchbook," Welles described how halfway through the broadcast "... we saw that there were a great many policemen, and every moment more .... I had no idea that I'd suddenly become a sort of ... national event." The dramatization had produced nationwide panic, with people believing the country had "fallen to the Martians." Many had tuned in late, missing the opening explanation, and because of previous frequent program interruptions due to the threat of war in Europe took the dramatization as a genuine news bulletin.

[16] Orson said that his mother taught him to read using Shakespeare's *A Midsummer Night's Dream* as a primer. He said, "I was marinated in poetry, and to learn right at the beginning, 'a sense of awe, delight, and wonder.' "

Worlds, interrupted by news of a reputed invasion from Mars that led to brief national panic across the United States.[15]

Dudley and his wife, Laura's, seminal roles in Orson's upbringing commenced following the jaundice-induced death of his mother, Beatrice, in 1924 when he was only nine years old.[16] Two months later, Laura took Orson and their three daughters east to "Sunset Farm" on Long Island near Syosset. Dudley then had long been committed to lead wealthy Chicago art patrons around Europe. Once Orson was alone

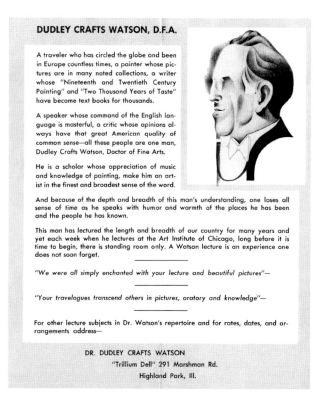

Dudley Crafts Watson, lecturer on travel and the fine arts. An encomium from one of his clients read, "Anecdotes of the world's great and small pour out with the apparent ease of a streamlined train streaking along at 110 miles an hour." Dudley's letterhead for the "1923 Caravan Deluxe" tour advertised "European Caravans, Music Picture Symphonies, and Lectures on Art."

Dudley Crafts Watson and daughter Marjorie. Her escapades with the young Orson Welles were the stuff of family legend.

with his cousins, they started playing doctors and nurses examining each other until Laura put a firm halt. Orson's extraordinary precociousness, which had labeled him as a verbal genius by age two, made it impossible for those in charge of him to punish him effectively, those close to him preferring to delight in his extraordinary ability to create artistic fantasies. Soon afterward, Orson went off to a math-strong grammar school in the university town of Madison, Wisconsin. The following summer of 1925, when both Dudley and Laura were in Europe, they sent their youngest daughter, Marjorie, with a governess to the effectively still classy Sheffield House Hotel of Grand Detour, Illinois. Then it had just been bought by Orson's father, Richard Hodgdon Head Welles (1872–1930), who had made his fortune with the invention of a carbide bicycle lamp; there Orson was to spend much of the summer. To stop being bored, he and Marjorie soon put on Orson's spare clothes and snuck out of the hotel to travel as itinerant actors over northern Illinois for three days, sleeping in the woods on pine needles and taking in a penny each time they found an audience for their artistic skills.[17]

In the fall of 1926 Orson began boarding at the extraordinary Todd School for Boys in Woodstock, Illinois, 50 miles to the northwest of Chicago. Not only did the Todd School have its own airport, but it was also a working farm with a large kennel that supplied dogs to neighboring homes. There boys from wealthy families prepared for East Coast colleges like Harvard and Princeton. Under its headmaster Roger

[17] Although the "official" version is that the State Rural Police of Illinois found the runaways in a small town, Marjorie's family maintains that the runaways were found on a street corner in Milwaukee, singing and dancing for money. Marjorie and Orson's friendship continued over the years, with Marjorie and her family often visiting Orson backstage at his theater productions.

A very precocious Orson at age 10.

[18] Roger Hill's daughter Joanne remembers that Orson "... was a very unusual boy. He had need of a very unusual education." When Orson entered the Todd School in 1926 at age 11 he was said to have scored 185 on the Stanford-Binet test. As Headmaster, Roger Hill had built a very progressive educational program at the Todd School that combined studies with activities such as flying instruction, working the school farm, and producing movies and plays.

[19] Roger Hill and Orson wrote the *Everybody's Shakespeare* series, first published by the Todd Press in 1934. These texts of three of Shakespeare's plays were "edited for reading and arranged for staging"—to make them "a living vibrant entity"—and were used by students and teachers throughout the United States. In a 1938 article by Welles and Hill in *The English Journal*, they maintained that Shakespeare's plays "... were written for you, for the groundlings, for the unscholarly Globe patrons who walked in from the cockfight on the street."

"Skipper" Hill, students were encouraged to move independently, choosing their own direction as opposed to those determined by their teachers. It was Todd's 200-seat theatre that most let Orson flourish at such a young age, he not only directing one play after another but playing most leading parts, say, that of Shakespeare's *Richard III*.[18,19]

After graduation in 1931, he balked at going off to college, persuading his mother Beatrice's former lover, his legal guardian Dr. Maurice Bernstein, to give him the cost of passage across the Atlantic to Ireland. Six months before, his never-confident, increasingly alcoholic father had died under mysterious circumstances in Chicago's famed Bismarck Hotel. Rumors abounded that Richard Welles committed suicide. Soon after his arrival in Dublin in the fall of 1931, Orson's self-confident demeanor and powerful voice led to his being given substantial acting roles (e.g., the Ghost and Fortinbras in *Hamlet* and Duke Karl Alexander in the tragic comedy from Lion Feuchtwanger's novel *Jew Süss*) for over three months at the artistically avant-garde Gate Theatre.[20] He was by then already a huge presence on the stage, already six feet tall and broad faced with much of his adolescent pudginess temporarily gone. When afterwards he moved back for a year to Chicago before finding unqualified success with John Houseman in New York

Orson at the Todd School.

Orson Welles's portrait in 1937, taken by Carl Van Vechten, famed portraitist photographer and writer, just before Welles burst on the national scene with his War of the Worlds *broadcast.*

City, Orson could never be sure what his oft well-intentioned but inherently unscrupulous guardian would let him do next. In contrast, Orson saw Dudley as someone wholeheartedly on his side.

In the early 1940s, Mother began working part-time at the housing office of the U of C. The monies so received helped her begin buying each year a pair of series tickets to let her and Dad hear famed soloists like Arthur Rubenstein and Arthur Schnabel play with the Chicago Symphony Orchestra at Orchestra Hall.[21] Increasingly then affecting Mother's and especially Dad's and my day-to-day morale was the constant stream of bad war news coming from Europe, where Hitler's armies went from victory to victory. England was barely hanging on with its survival being threatened by the growing packs of German submarines. Dad thought for England to finally prevail the United States had to join the war. But the strong Midwestern isolationist movement personified by Colonel Robert McCormick's *Chicago Tribune* newspaper kept Franklin Roosevelt from asking our Congress to declare war. Happily, the *Chicago Tribune* morning monopoly was broken by Marshall Field's new *Chicago Sun*, whose first issue appeared on December 4, 1941, just before the devastating Japanese attack on Pearl Harbor that destroyed most of our Pacific Fleet battleships. Many more months of bad news from Southeast Asia further lowered our morale until our breaking of the Japanese secret naval codes led to defeat of the Japanese Navy at the June 1942 battle of Midway. The subsequent German debacle of Stalingrad and their surrender on February 2, 1943, finally raised hopes that victory, though likely years away, would inevitably come.

[20] Actor Micheál MacLiammóir, co-founder of The Gate, famously described the Orson that showed up in Dublin in his book *All for Hecuba*—"He knew that he was precisely what he himself would have chosen to be had God consulted him on the subject of his birth."

[21] "Somehow or other I managed to fight off reading *Gone With the Wind*. After reading some of the reviews of the moving picture I hope that I'll be equally successful in not seeing the movie." —JDW, Sr., January 30, 1940, diary entry

CHAPTER 5

Ideas (Great Books) over Facts (Textbooks) (1943–1952)

JDW, Sr.

Soon after I started my 1942–1943 sophomore year at South Shore High School, Mother and Dad began dreaming that it might be my last year there. They wanted me to move on to a brand-new program at the University of Chicago (U of C) that admitted qualified students after two years of high school. Though I was just one of several bright South Shore students, I was its most voracious reader of books and would most profit by being accepted into its second class. Our family, however, could not cover the tuition of 100 dollars a quarter and my going there depended on my being awarded a full tuition scholarship. After the scholarship exam, I was selected to come in for an interview that let me talk about the books I had recently read. In later being awarded my scholarship, I must have been aided by mother's growing friendship with Valerie Wickham, the Director of Admissions, whom she first met through her Housing Office connections. Soon my task would be to show that I intellectually belonged on the U of C "Midway" campus.

The U of C then long had the reputation of being the Midwest's, if not our nation's, foremost academic research powerhouse. Being at the forefront of research as well as undergraduate teaching was seen as essential to its existence by its founder, the extraordinarily gifted Baptist preacher William Rainey Harper. Born in 1856 into a Scotch Presbyterian covenanter family living in a log cabin in New Concord, Ohio, Harper was highly precocious, learning to read while three and entering Muskingum College at age 10 and graduating at 14. After spending sev-

William Rainey Harper (1856–1906). In addition to his educational endeavors, Harper was the author of many books on religion, languages, and education.

[1] The first University of Chicago had been founded in 1856 but was bankrupt in 1886. Rockefeller was approached by a trustee of the defunct school, and he agreed to donate $600,000 if other wealthy Chicagoans would donate $400,000.

[2] Classes began at the U of C in October of 1892 with almost 600 students and 103 faculty members.

eral years working in the family store, he went to Yale at 17 for graduate work. Quickly he mastered Greek and Latin, going on to Hebrew, earning his Ph.D. in two years, and graduating just before his 19th birthday. Soon after marrying Ella Paul, he spent a year as head of Masonic College in Macon, Tennessee, and then a year at Denison University in Granville, Ohio. At only the age of 22, in 1878 he began teaching Hebrew at the Baptist Union Theological Seminary in Chicago, where he stayed for six years before returning to Yale in 1884, teaching Hebrew in the Semitic Language Department and the Divinity School.

When the American Baptist Educational Society decided to form a Midwestern university, Harper joined the organizational committee and with John D. Rockefeller promising a million dollars of support was offered its presidency in February 1891, taking office as president of the new U of C on July 1, 1891. Though the U of C was initially envisioned as a Baptist undergraduate college, Harper had bigger ideas, wanting to promote his ideal of higher education that combined strong undergraduate teaching with a strong support of research. Toward that end, he strove to recruit first-class faculty through high salaries made possible by further fundraising that included another million dollar pledge from J.D. Rockefeller. Through all his administrative actions he continued to produce biblical commentaries while maintaining a passionate interest in sports that soon made the new U of C also a Big Ten sports powerhouse.[1,2]

In the fall of 1901, while then still only 46, Harper presided over a biennial celebration that included a wealth of conferences, sermons, lectures, and a string of faculty publications that extended to 25 volumes.[3] Cornerstones were laid for six buildings and one of the new women's dormitories was dedicated. Tragically, he was to live for only four more years with his doctors telling him they had found cancer early in 1905. Resolutely he continued to write, teach, and confer with colleagues until shortly before his death in January 1906.

Newspapers had a field day with Harper's efforts to raise money for the university, especially from Rockefeller, as here in this Ralph Wilder cartoon. At the time Rockefeller said that "I confidently expect that we will add funds from time to time to those already pledged..." and he was good for his word. Through his foundations he eventually donated some $80 million to the University.

[3] **Harper envisioned the University's goal as creating scholar–researchers. He paid the faculty well, reduced their teaching loads, and championed coeducation. Notable early faculty members included John Dewey, Jacques Loeb, and Amos Alonzo Stagg. Harper also established the University of Chicago Press.**

Happily, the presidents that followed him over the course of the next 23 years—Harry Pratt Judson, Ernest DeWitt Burton, and Max Mason—maintained his vision for the U of C being exceptionally strong in research. In a 1925 ranking of its graduate departments, as determined by polls of leading research academics, eight of its departments ranked number one in research—Botany, French, Geography, Geology, Mathematics, Physics, Sociology, and Spanish. The University was number two in Astronomy, Education, Political Science, and Zoology; with Economics, History, Philosophy, and Psychology ranking third;

[4] Rosenwald made his fortune as a partner in Sears, Roebuck and Co. His philanthropy extended beyond the U of C and his funding of the Museum of Science and Industry in Chicago. He established 5,000 or so "Rosenwald Schools" in the South to educate poor rural black youth as well as some 4,000 libraries, and he championed the efforts of Booker T. Washington, endowing his Tuskegee Institute and supporting the early civil rights movement. His article in the May 1929 *The Atlantic Monthly*, "Principles of Public Giving," laid out his philosophy of philanthropy—not giving in perpetuity, no profligate spending of the principal, and the value of ideas as well as money.

[5] In an odd footnote to Chapter 3 of this book, Rosenwald's grandson Armand Deutsch believed that he was the original murder target of Nathan Leopold and Dickie Loeb. Loeb's father was a Sears vice president, and the families knew each other. Instead of walking home the day of the Franks murder, Deutsch had been picked up by the family chauffeur to be taken to a dental appointment. If he had been walking home from school that day, Deutsch later said "I surely would have accepted [the] invitation for a ride."

and Chemistry and English fourth. Only German was in fifth position. No other institution, even those at the top of the Ivy League, came even close to its rankings.

The large proportion of university funds being so devoted to research came, however, at the expense of its undergraduates, whose tuition monies were ever-increasingly devoted to its graduate program. By the end of Max Mason's brief presidency, considerable progress had already been taken by the U of C administration to strengthen undergraduate education through the development of several new year-long survey courses to be taken during the first two years of undergraduate education. In a 1928 report, Dean Chauncey S. Boucher urged the abandonment of four different undergraduate programs. In their place, the report proposed that the College of the University of Chicago develop general introductory "survey" courses for the first two years of all undergraduate life. "Electives" would then dominate the junior and senior years. Toward support of the proposed new undergraduate initiatives, Julius Rosenwald offered $2 million for a collection of new dormitories across the Midway from the main campus. Today they bear the names of the second and third presidents, Judson and Burton.[4,5]

The early 1929 choosing of Robert Maynard Hutchins, then Dean of Yale's Law School, as the U of C fifth president was also seen as a means to strengthen undergraduate teaching. Dad was very much a Hutchins partisan, remembering him well from 1915–1916 when they were close friends as freshmen at Oberlin College. There Bob's preacher father Will taught the obligatory Bible class that had first made Dad doubtful of the existence of God. After wartime ambulance driving service in the Italian mountains, Hutchins transferred to Yale in the fall of 1919 for his last two undergraduate years. There he equally excelled at debating, effectively arguing that the only proper way to move forward was through the truth. After a year teaching at a preparatory school in Lake Placid, New York, he returned to Yale to serve as secretary to its new president,

James Rowland Angell, while simultaneously attending its Law School, from which he received the LLB *magna cum laude* in 1925. That fall he began teaching law, becoming Yale's acting Dean in 1927 and Full Professor and Dean in early 1928, when he was only 28 years old. Like many if not most young academics of the immediate post-WWI period, he never had focused his thinking exclusively on the Bible, accepting new profundities arising from the social sciences. Teaching, say, sociology, to them was a step upward from mere preaching from the Bible. It was based on reason and so helped liberate society from outworn beliefs. Increasingly, salvation meant accepting truth gained from experiencing today's life. Still much affecting Hutchins's teaching, however, remained the Evangelical Protestant culture of his father and Oberlin. Then the paramount virtues were courage, temperance, liberality, honor, justice, wisdom, reason, and understanding—virtues he had accepted from his father and the Oberlin community. Of equal importance were the philosophical writings that always moved him—Plato's *Dialogues*, Aristotle's *Ethics* and *Politics*, and John Stuart Mill's work. Strong affirmations of rational thought and discussion became a major theme for all of Hutchins's subsequent educational proposals.

Julius Rosenwald (1862–1932) and educator Booker T. Washington (1856–1915).

Though Yale's president James R. Angell, who had been recruited to the U of C faculty in 1895 by John Dewey and later rose to be Dean of its Faculty in 1908, warned the U of C trustees that Hutchins needed more time to mature before taking on the responsibility of a great university president, they did not accept his advice. Hutchins's charismatic, idea-driven personality gave him the potential to be a second William Rainey Harper. By the spring of 1929, he was already in Chicago as the U of C's fifth president.[6]

Hutchins's most innovative as well as controversial action as president between 1929 and 1951 came from extending Chicago's just-introduced broad freshman and sophomore survey courses into the last two years of high school. Then he regarded the junior and senior years

[6] **During the course of his administration, Hutchins eliminated varsity football and organized the University's graduate departments into four academic divisions—biological sciences, humanities, physical sciences, and social sciences. But he also invited controversy with his establishment of the medical school at the University and many of his administrative and educational ideas, including imposing a "particular philosophy" on the University, his "socialist" proposal to abolish distinctions of faculty rank, and his perceived antiscientific stance.**

of American high schools as too often wasted. They were not using the mental capacities of our nation's brighter students, who by 16 years should be learning to think logically as opposed to memorizing facts in the absence of ideas that make sense of them. Therefore, Hutchins's four-year college replaced the conventional textbooks of the U of C first survey courses with the Great Books of Western Civilization, like *The Wealth of Nations* by Adam Smith or *The Acquisitive Society* by R.H. Tawney, Hutchins believing that their collective ideas best provided the intellectual foundations for modern life.[7]

Robert Maynard Hutchins (1899–1977), the "boy" president. In molding the "University of Utopia," he believed that "The purpose of the university is nothing less than to procure a moral, intellectual, and spiritual revolution throughout the world."

Robert Hutchins's choice of the Great Books as the primary vehicle for academia's searches for profound human truths grew out of his already many-year-long intellectual partnership with the New York–born and Columbia-educated young philosopher, Mortimer J. Adler. Adler became first enamored by the teaching potential of the Great Books in 1921–1922 when Hutchins had just started pursuing his Yale law degree, through participating in a post-WWI "Great Books" course at Columbia College taught by the well-regarded Professor of English, John Erskine. Originally designed for Army recruits idled for some months in Europe at WWI's end, it became available to Columbia students during Adler's junior year under the rubric "General Honors." Most excitingly, it met for 60 two-hour sessions throughout two academic years to study one class of Western thought a week—beginning with Homer, Herodotus, and Thucydides and ending with Darwin, Marx, and Freud.

[7] Although often labeled as antiscientific, Hutchins's position on science was much more nuanced—that "there are other means of obtaining knowledge than scientific experimentation." In his *Great Books: The Foundation of a Liberal Education* (1954), he wrote "… faith in experimentation as an exclusive method is a modern manifestation… it is now regarded in some quarters … as the sole method of obtaining knowledge of any kind." He went on to say that "it is necessary to think" to solve a problem and the "experimental scientist cannot avoid being a liberal artist" as "men of imagination and theory as well as patient observers of particular facts…."

Though Adler never received his bachelor's degree, having refused to participate in physical education, he never expected to be refused admission to Columbia's Ph.D. program in Philosophy.[8] His continuous belittling of pragmatism in favor of metaphysics at department gatherings, however, had made its Philosophy faculty more than occasionally allergic to his sight or sound. After his acceptance, instead, into Columbia's graduate program in Psychology, Adler, for his undergraduate teaching requirements, co-led his own "general honors Great Books seminar" with the equally talented, young New York literary intellectual Mark Van Doren.[9] As the course developed over the next few years, Adler continually reconstructed the list of classics that their course would discuss. Much favoring Adler's ever-growing Great Books obsession was his growing friendship with the young American philosopher Scott Buchanan, who after a Rhodes Scholarship at Oxford had moved to Harvard, where he obtained his Ph.D. in Philosophy. Later Buchanan became Dean of St. John's College in Annapolis, Maryland, and established the Great Books program there as the centerpiece of the College's curriculum.

Robert Hutchins's first knowledge of Adler's passion for Great Books occurred after they met when he was still Yale Law School faculty. Quickly Hutchins admired Adler's familiarity with philosophy and prodigious knowledge of the literature of psychology. After becoming president of the U of C, Hutchins for more than a year unsuccessfully sought a position for Adler in his Philosophy Department. Given his metaphysical bent, however, Adler had no place in a department that much preferred pragmatism over metaphysics. Intrinsically, they had little sympathy for hosting an Adler-led community of young philosophers dedicated to a dialectic examination of contemporary academic knowledge disciplines. Adler's social and intellectual exchanges with fellow academics, however, revealed total disdain for the methodologies of social science. In contrast, he believed that philosophy was rational science as opposed to empirical science. As Adler's arrogantly expressed

Mortimer J. Adler.

[8] Adler refused to take the required swimming test. Degreeless, Adler nonetheless applied to be an instructor in Psychology in 1923 and earned his Ph.D. in 1928. Columbia finally relented in 1983, awarding him an honorary bachelor's degree that year.

[9] Adler and Van Doren, along with Buchanan, also taught the Great Books course at New York's Cooper Union in classes aimed at immigrants and laborers.

[10] Although the Great Books were meant to give students an education in the liberal arts—according to Adler "the basic skills of learning"—the program was often criticized as being Eurocentric for its lack of representation of Asian, minority, and female authors. Its philosophy pitted supporters of cross-disciplinary learning (Hutchins et al.) versus pragmatists, such as John Dewey and Sidney Hook.

[11] Hutchins asked the University Law School to hire Adler in 1930 as Professor of the Philosophy of Law. He became a full professor in 1942. Adler resigned in 1952 to found the Institute for Philosophical Research in San Francisco, which he then moved to Chicago in 1963.

[12] Adler believed that the "universal truths" propounded by such as Plato, Aristotle, and especially St. Thomas Aquinas are "what constitutes a good education for all men at all times and places simply because they are men." Adler was a prolific writer and authored as many as 50 books that drove these points home. His 1954 television series *The Great Ideas* reached a broader audience during the fledgling years of television's Golden Age.

unwanted opinions increasingly became widely known, Hutchins reluctantly gave up trying to get him a position in the U of C Philosophy Department. Instead, he placed him in its Law School as well as making him a member of the College faculty. Thus segregated, his pro–Great Books views would not necessarily generate further faculty hostility.[10,11]

Most important in the reshaping of Hutchins's mind during his early years as the U of C fifth president from the methodologies of Law and the Social Sciences to the Great Books was his weekly co-participation with Adler in a 20-student seminar that met once a week for a two-hour seminar on the "great book" of the week. As the class ran over two years, course credit and examinations were suspended. At each year's end, students were orally tested by clever outsiders like Scott Buchanan. Over time, Hutchins and Adler extended the course material down to University High School or up to the Law School using a core group of newly appointed faculty serving as its teachers. Teaching with the Great Books presented major pedagogical challenges, with Adler's rude badgering of students necessarily counterbalanced by Hutchins's more humorous bantering. For both individuals, the predominant pedagogical methods were discussions about words and ideas.[12]

Sensing that Hutchins's Great Books undergraduate curriculum would further reduce the role of the university's departments like English, History, and Physics in undergraduate education, many of the U of C's most prominent faculty members began to seriously contest Hutchins's ideas. Moreover, Hutchins, after unsuccessfully wanting to be chosen as FDR's Vice President in 1940, soon after had emerged as one of our nation's most vocal isolationists, strongly opposed to giving military assistance to an England almost on the ropes. Ever increasingly, the U of C faculty saw Hutchins's presidency as a disaster that would only end by his resignation.

His major educational initiative, the "four year" college that started after two years of high school, would never have come into existence in 1942 were it not for the departure of many key faculty members to wartime duties. Only by a single vote in April 1942 did the faculty re-affirm an earlier January 1942 vote to confer a bachelor's degree upon the completion of the college degree. Not stopped by the narrowness of his victory, Hutchins quickly assembled a college faculty whose primary role was to *teach*, as opposed to turning out academic research.

My first classes were in June (not September) 1943, reflecting a perceived need to move males through the college faster than would have occurred were there no war. All too soon, we were likely to have our education interrupted by military service.

Soon after Pearl Harbor and the Declaration of War on the United States by Germany, Mother took on a full-time job in the Chicago Loop becoming Director of Personnel for the Chicago branch of the American Red Cross. All of my uncles as well as Dad were by then too old to be wanted for military service. For a few weeks after Pearl Harbor, Dad joined a group of similarly aged men doing military drills in Stagg Field, once the site of the U of C glory days in football. All too soon it

Jim Watson during his student years (1943–1947) at the U of C.

JDW, Sr. (at right), at Stagg Field. Stagg Field was also the site (under the west stands) of top-secret Chicago Pile-1, where the first self-sustaining nuclear chain reaction occurred on December 2, 1942.

became clear that our nation had better ways for them to serve the war effort. Soon finding a well-paying job opened up by its prior occupant's departure for military service was Uncle Tom. His family finances somehow flourished much more in the war than before or afterward. How close he was to activities that soon became known as the "black market" Dad did not want to know.

Dad's brother Bill, then at Yale on its Physics faculty, returned to Chicago in mid-1943 to help run a top-secret atomic physics effort, the Manhattan Project, by then still centered at the U of C's Metallurgical Laboratory—the "Met Lab." Coming with Uncle Bill to Chicago was his family, wife Betty and daughters Ruth and Alice, who went to the Laboratory School only several blocks away, where I had gone 10 years before to its Nursery School. Bill was strictly forbidden to say what he was up to, only letting us guess that something really big was up through mentioning his coworkers going back and forth to Tennessee and Washington State.[13] During the winter, Uncle Dudley invited Bill and Betty to

[13] The Met Lab was initially focused on developing chain-reacting "piles" to extract plutonium from irradiated uranium and had produced the first weighable amount of plutonium in August of 1942. The U.S. Military had come to Hutchins, who was at the time a professed isolationist, in 1941 to take on the prime responsibility for the Met Lab. He was said to be the only non-scientist outside the government who knew about the project. Hutchins later said that, in hindsight, he would not have agreed to having the project at the U of C.

The Watson brothers—Stanley, Thomas, William, and James.

the Art Institute on the occasion of Orson Welles coming to his office to get Dudley's approval of his new wife, Rita Hayworth. Her face on the cover of *Life* magazine had so caught Orson's fancy that he immediately wanted to marry her. Bill and Betty found Rita nervously likeable, obviously uncomfortable at seemingly needing Dudley's approval even after Orson had tied the knot with her.

Only after the war ended was my growing hunch confirmed that the Metallurgical Lab's purpose was to build uranium-based bombs many orders of magnitude more powerful than conventional explosives. The project had its origins in a by-now-famous letter composed by Hungarian-born theoretical physicists Eugene Wigner and Leo Szilard, written for Albert Einstein to send in July 1939 to President Franklin D. Roosevelt, calling for the United States to develop atomic weapons before Nazi Germany did. After the December 7, 1941, Japanese attack on Pearl Harbor, Arthur Holly Compton, the Nobel Prize–winning head of the U of C Physics Department and a long close friend of Uncle Bill, consolidated the resulting American nuclear research at the U of C, with the goal of producing the first atom bomb by January 1, 1945.[14] Enrico Fermi and Leo Szilard, after moving from Columbia University to Chicago, oversaw the building of the uranium and graphite piles, Chicago Pile-1,

*Orson and Rita were married on September 7, 1943. Although Orson told her that "You are my life—my very life," things did not work out for them, and they divorced in 1948. Rita famously offered that, "I can't stand his genius anymore." They made one movie together—*The Lady from Shanghai—*in 1947.*

[14] Arthur Holly Compton (1892–1962) won the 1927 Nobel Prize in Physics for showing the particle concept of electromagnetic radiation—the "Compton effect." He had been Professor of Physics at the U of C since 1923, studying X rays and cosmic rays. He served as director of the Met Lab from 1942 to 1945. After the successful run of Chicago Pile-1, Compton called James B. Conant, chairman of the National Defense Research Committee, to say "... the Italian navigator has just landed in the new world," to which Conant replied, "Were the natives friendly?" Compton assured him that, "Everyone landed safe and happy." After the war, he ended his career at Washington University as Chancellor and Professor of Physics.

Compton's Manhattan Project–related identification badge for the Hanford, Washington, plutonium-generating site, complete with fake name for security.

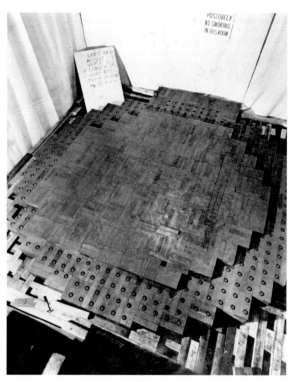

Chicago Pile-1 under construction showing the 19th layer of graphite being added. The pile was constructed in 16 days of alternating layers of blocks of graphite and graphite containing uranium metal and uranium oxide. In all, 57 layers were constructed to reach criticality. The pile measured 8 feet × 8 feet at the bottom and reached a height of 16 feet. It was controlled by cadmium rods that acted as neutron sponges.

[15] Enrico Fermi (1901–1954) won his Nobel Prize in 1938 for his work on neutron physics. He had been at Columbia University from 1939 to 1942 after fleeing Fascist Italy, using his Nobel trip to Stockholm as the means of escape. He built his initial graphite and uranium oxide pile at Columbia with Szilard and then was moved to the Chicago Met Lab, where he built Chicago Pile-1. Szilard and Fermi were issued a U.S. patent for the pile after the war.

[16] Leo Szilard (1898–1964) had conceived the idea of a nuclear chain reaction in 1933 in London while waiting for a red light to change. He had also read H.G. Wells's 1914 novel *The World Set Free*, which described nascent ideas of "atomic bombs." A refugee of Nazi persecution, he fled first to London, in 1938 to Columbia University, and then with Fermi to the U of C's Met Lab. An active campaigner against the actual use of the atomic bomb against Japan, he turned to biology after the war, and was a frequent visitor to Cold Spring Harbor. Szilard was instrumental in the creation of the Salk Institute for Biological Studies.

that were finally brought to a critical mass for a sustained nuclear chain reaction on December 2, 1942.[15,16]

By the time I arrived on the U of C campus in June 1943 as a college student, the first pile was already abandoned, with subsequent industrial-level pile development transferred to more appropriate sites in Hanford, Washington, and Oak Ridge, Tennessee. The initial formulation for actual bomb designs also started on the U of C campus in top-secret Eckhart Hall physics laboratory space. By early 1943, it moved to Los

Alamos, New Mexico, under the direction of the University of California, Berkeley, theoretical physicist, J. Robert Oppenheimer.

Uncle Bill's Chicago stay lasted only a year, after which he and his family moved to Montreal to let him become the American liaison officer to a newly formed UK–Canadian–French atomic physics venture.[17] Under the Quebec Agreement of 1943, attended by FDR and Churchill, a heavy water pile was to be built in Canada largely staffed by English physicists engaged until then in defeating the German submarine blockage of the United Kingdom. Bill reported directly to General Leslie Groves, the military director of the Manhattan Project. Groves did not trust the British-led team because most of its members were not British. In addition, he did not think that the Canadian project would be of much

Fermi's Los Alamos security badge photo.

Leo Szilard. Szilard's concept of critical mass was essential to producing a self-sustaining chain reaction. But he also realized the consequences of this, and when the chain reaction was achieved in Chicago he later described it as a "black day in the history of mankind." Szilard was a founding sponsor of the Bulletin of the Atomic Scientists *in 1945 and after WWII campaigned to curb the nuclear arms race.*

[17] **William Weldon Watson (1899–1992) served as Division Director of the Met Lab for a year starting in September 1943 before moving to the Canadian heavy water project. His area of research was the process of separating differing isotopes of atoms, critical to the development of the atomic bomb. He had started his work on the atomic bomb project at Yale before the Secretary of War sent him to Chicago. He was a product of the U of C, having earned all of his degrees there and serving on its faculty until he moved to Yale in 1929. He returned to Yale in September of 1945 after two atomic bombs had been dropped on Hiroshima and Nagasaki in August of 1945, and remained there until 1968. He was one of the first scientists to publicly support using nuclear energy for powering aircraft and space satellites.**

[18] Ending WWII was an incremental process. The Italian Fascist government collapsed in July 1943, and Italy approved a conditional surrender on September 8, 1943; the Germans and the Allies both invaded Italy, and the Allies could not claim victory there until the spring of 1945. The June 6, 1944 D-Day landing of the Allies in Normandy was the beginning of the end for the German forces, and their surrender came in May of 1945. The Pacific war against Japan ended with their formal surrender on September 2, 1945.

help to the war because the plant at Hanford, Washington, would be able to supply enough fissionable uranium products for the war effort. The Americans were to provide uranium and heavy water for the Montreal project, but there was to be no information provided to the British-led team on the chemical separation of the fissionable plutonium. As soon as the war ended, Uncle Bill left Montreal to be back at Yale by the start of the 1945–1946 school year.[18]

During the months following the war's end, Robert Hutchins, to the delight of his faculty scientists, moved quickly to solidify the U of C's leading position in Physics. By appointing universally admired, Italian-born Enrico Fermi to its physics staff, he ensured a leading role in postwar physics for the U of C. At the same time, Hutchins approved setting up several new research institutions, including a new Institute of Nuclear Studies, which not only Enrico Fermi would join but also the immensely clever Hungarian Edward Teller. Leo Szilard would, in turn, become a founding member of a new Institute of Biophysics to let him move his primary interests from Physics to Biology. By so strengthening science and becoming a spokesman for the future responsible use of atomic energy, Hutchins had suddenly become much less controversial, with calls for his resignation much less frequently heard.[19]

Hutchins and Adler saw undergraduate liberal education exposure to the Great Books as but the start of a lifelong examination of the fundamental theories and practical problems that have confronted mankind as a social animal. Both reluctantly came to realize that adults, as opposed to youths, are more deeply drawn to ideas and their relevance to the great problems of human life. Long before the war ended, Adler had withdrawn from teaching undergraduates to lead adult education Great Books programs at the U of C downtown university college campus.

Mother and Dad became part of one of the Chicago region's first Great Books adult education efforts, joining a group effectively led by the intellectually inclined Hyde Park dentist Al Dahlberg, whose still

Date of Meeting	Book	Leader
Sat. Dec. 8th	The Acquisitive Society - Tawney	Jean Watson
Sat. Jan. 12th	Dead Souls - Gogol	George Bock
Sat. Feb. 9th	Anthropology in Modern Life - Boaz	Thelma Dahlberg
Sat. March 9th	Alice in Wonderland - Carroll	Vida Wentz
Sat. Apr. 13th	Common Law - Holmes	Barbara Bock
Sat. May 11th	Das Kapital and Manifesto - Marx	Lee Wentz
Sat. June 8th	Decline of The West - Spengler	Harry St. John
Sat. July 13th	Private Papers of Henry Rycroft - Gissing	Morrie Moran
Sat. Aug. 10th	King Lear - Shakespeare	Frances St. John
Sat. Sept.14th	Comedies - Moliere	Willie Moran
Sat. Oct. 12th	The Social Contract - Rousseau	Jim Watson
Sat. Nov. 9th	War and Peace - Tolstoi	Al Dahlberg

Schedule for the Jean and James Watson's Great Books group.

very young sons Jim and Al, Jr., would go on like me to serious careers as molecular biologists at the University of Wisconsin and Brown University, respectively. Meeting in its members' houses once a month on Saturday evenings, the group's 1945 discussion of December 8 was led by Mother on *The Acquisitive Society* by the highly influential Englishman Richard Henry Tawney. In our South Shore neighborhood, a Great Books discussion reading club soon after came into existence at the Bryn Mawr Community Church located at 7000 Jeffrey Avenue.

To widen further public exposure to the Great Books, Hutchins increasingly used the resources of a much-revamped *Encyclopedia Britannica* (EB), which its former owner Sears, Roebuck had effectively put under control of the U of C. Masterminding the transaction was his close friend William Burnett Benton, who came to the U of C in 1937 as a half-time vice president. A founder of the New York–located ad-

[19] Hutchins, after having put the world in what he later said was "moral jeopardy" with the Manhattan Project research at the U of C, was concerned with how atomic energy could be controlled for peacetime use. He was instrumental in the formation of the Federation of American Scientists in 1945, which was an amalgamation of individual atomic scientist groups formed as a result of the Manhattan Project and whose initial aim was to prevent nuclear war. Hutchins had given the Chicago group, Atomic Scientists of Chicago, $10,000 to support this mission. He also gave substantial support to the passage of the McMahon Act (Atomic Energy Act of 1946), which turned over management of the development of nuclear weapons and nuclear power to civilian, rather than military, control.

```
                                    GREAT BOOKS


       Homer:  The Iliad; The Odyssey

       Aeschylus:  Tragedies

       Sophocles:  Tragedies

       Euripides:  Tragedies

       Aristophanes:  Comedies

       Herodotus:  History of the Persian War

       Thucydides:  History of the Peloponnesian War

       Hippocrates:  Collection on Medical Writings

       Plato:  Dialogues, especially The Republic; Symposium; Phaedo; Meno; Apology

       Aristotle:  Works, especially Ethics; Poetics; Politics

       Euclid:  Elements of Geometry

       Cicero:  Works, especially Orations; Offices; Tusculum Disputations

       Lucretius:  Of the Nature of Things

       Virgil:  Aeneid

       Horace:  Odes and Epodes; The Art of Poetry

       Livy:  The History of Rome

       Ovid:  Metamorphoses

       Quintilian:  Institutes of Oratory

       Plutarch:  Lives

       Plotinus:  Works

       Tacitus:  Dialogue on Oratory; Germania; Agricola

       Nichomachus:  Introduction to Arithmetic

       Epictetus:  Discourses

       Lucian:  Works, especially The Way to Write History and The True History

       Marcus Aurelius:  Meditations

       Galen:  Of the Natural Faculties

       The Bible

       Augustine:  Confessions; The City of God
```

A list of the Great Books from the mid-1940s. (Figure continued on following pages.)

-2-

Volsunga Saga

Song of Roland

Burnt Njal

Maimonides: Guide for the Perplexed

Aquinas: Summa Theologica

Dante: Divine Comedy

Chaucer: The Canterbury Tales

Thomas A. Kempis: Of the Imitation of Christ

Leonardo Da Vinci: Notebooks

Machiavelli: The Prince

Erasmus: The Praise of Folly

More: Utopia

Rabelais: Gargantua and Pantagruel

Calvin: Institutes of the Christian Religion

Montaigne: Essays

Cervantes: Don Quixote

Edmund Spenser: The Faerie Queene

Francis Bacon: The Advancement of Learning; Novum Organum

Shakespeare: Plays

Galileo: Dialogues Concerning Two New Sciences

Harvey: On the Motion of the Heart

Grotius: The Law of War and Peace

Hobbes: Leviathan; A Dialogue of The Common Laws

Descartes: A Discourse on Method; Principles of Philosophy; Meditations

Corneille: Tragedies, especially The Cid; Cinna

Milton: Paradise Lost; Areopagitica

Moliere: Comedies

Spinoza: Ethics; Political Treatises; Short Treatise on God, Man and his Well-Being

-3-

Locke: An Essay Concerning Human Understanding

Racine: Tragedies

Newton: Mathematical Principles of Natural Philosophy; Opticks

Leibnitz: Discourse on Metaphysics; New Essays Concerning Human Understanding

Defoe: Robinson Crusoe; Moll Flanders

Swift: Gulliver's Travels; Battle of the Books

Montesquieu: Persian Letters; Spirit of Laws

Voltaire: Candide; Toleration; Letters

Berkeley: A New Theory of Vision; The Principles of Human Knowledge

Fielding: Tom Jones; Joseph Andrews

Hume: Enquiry Concerning Human Understanding

Rousseau: The Social Contract; Discourse on Inequality

Sterne: Tristram Shandy

Adam Smith: The Wealth of Nations

Blackstone: Commentaries on the Laws of England

Kant: Critique of Pure Reason; Prolegomena to any Future Metaphysics

Gibbon: The Decline and Fall of the Roman Empire

Hamilton and Jefferson: The Federalist Papers (along with The Articles of Con-
federation; the Constitution of the United States; and the
Declaration of Independence)

Bentham: Introduction to the Principles of Morals and Legislation; Theory of
Fictions

Goethe: Faust

Ricardo: The Principles of Political Economy and Taxation

Malthus: An Essay on the Principles of Population

Dalton: A New System of Chemical Philosophy

Hegel: Science of Logic; Philosophy of Right; Philosophy of History

Guizot: History of Civilization in France

Faraday: Experimental Researches in Electricity

Comte: Positive Philosophy

A list of the Great Books continued
from previous page.

-4-

Balzac: Works

Lyell: The Antiquity of Man

Mill: Utilitarianism; On Liberty; System of Logic

Darwin: On the Origin of Species

Thackeray: Vanity Fair; Henry Esmond

Dickens: Works

Claude Bernard: Introduction to Experimental Medicine

Boole: Laws of Thought

Marx: Capital; The Communist Manifesto

Melville: Moby Dick

Dostoevski: Crime and Punishment; The Brothers Karamazov

Buckle: A History of Civilization in England

Flaubert: Madame Bovary

Galton: Inquiries into Human Faculty and its Development

Ibsen: Plays

Tolstoi: War and Peace; Anna Karenina

Wundt: Physiological Psychology; Outline of Psychology

Mark Twain: Adventures of Huckleberry Finn; A Connecticut Yankee in King Arthur's
            Court

Henry Adams: The Education of Henry Adams

Charles Peirce: Chance, Love, and Logic

Oliver W. Holmes: The Common Law; Collected Legal Papers

William James: Principles of Psychology; The Varieties of Religious Experience

Poincare: The Foundations of Science

Freud: Outline of Psychoanalysis; Civilization and its Discontents

[20] **William Burnett Benton (1900–1973) was, like Hutchins, the son of a clergyman. He first met Hutchins when both were students at Yale. His varied career included life as the adman who introduced consumer surveys and the singing commercial to the world of advertising, vice president of the U of C (1937–1945), ownership of the Muzak Corporation (1939–1940), an Assistant Secretary of State (1945–1947), and U.S. Senator from Connecticut (1949–1953). Benton bought *Encyclopedia Britannica* from Sears, Roebuck and Co. in 1943 for $100,000 and donated a two-thirds beneficiary interest to the U of C (the university is said to have received some $60 million in royalties from this donation between 1943 and 1980). He served as Chairman of the Board and Publisher from 1943 to 1973.**

vertising agency Benton & Bowles, he was an advertising world genius whose intellectual interests were seemingly unlimited. Just after the Japanese surrender, he organized a Fat Man's Book Club for intellectually ambitious Chicago businessmen.[20] They met at the University Club in the Loop to discuss the Great Books with Huchins and Adler. Although Hutchins was wildly optimistic when he forecast an audience of 15 million adults for Great Books discussion groups, the EB-sponsored *Great Books of the Western World* collection was to grow into sets of 54 individual books, collectively selling nearly 1 million volumes. In the introductory volume *The Great Conversation*, Hutchins argued that beginning with the Greeks the "emergence of democracy as an ideal" was traceable to a liberal form of education.

William Benton, Robert Hutchins, and Mortimer Adler with the Great Books.

Once my sister Betty also acquired a scholarship and could enroll at the U of C, Dad and Mother seriously considered selling our South Shore 7922 Luella Avenue home.[21] By then LaSalle Extension University had moved from its increasingly crime-riddled, near-Southside neighborhood to a building within the Loop on Dearborn Street. Dad's now-improved salary, when added to monies that Mother would get through working more hours at the Housing Office, led them to briefly consider buying a 1900s wooden house on Harper Avenue, almost adjacent to the 56th Street Illinois Central Railroad station that Dad would use if they were to move. Then, after thinking more carefully of their potentially tightened financial state if Mother's employment at the U of C did not become full-time, they resigned themselves to more years of South Shore residence.

I was by then starting to grow toward six feet and had acquired the self-confidence to go on many hour-long birding trips by myself to the swamps bordering Wolf and Calumet Lakes. Then I could use my new postwar Bausch & Lomb 7 × 50 binoculars, which I purchased using money I earned from three 1942 *Quiz Kids* radio appearances. One spring day along Lake Calumet I saw both the Wilson and the Red Phalarope. Piping Plovers still nested on the sandy shores of Wolf Lake, which in early fall attracted large flocks of both Red-Breasted Sandpipers and Golden and Black-Bellied Plovers. Likewise Dad stopped always being with me when I boarded South Shore and South Bend trains to bird in the Indiana Dunes State Park at Tremont, some seven miles to the west of Michigan City. Prairie Warblers then nested close to its lakeshore high sandy dunes, with my always counting on hearing the buzz sounds of the Blue-Winged Warblers that nested in the more-inland swampy woods.

Just within the park on the road coming from the train station was a modest wooden hotel that dated to before the park was established. In the week before the end of my last exam and the U of C June 1947

Elizabeth "Betty" Watson as a student at the U of C.

[21] In a small-world coincidence, molecular biologist Matt Meselson, who graduated from the U of C in 1951, knew Jean Watson and dated Betty Watson before he knew JDW, Jr. Meselson played a critical role in the story of DNA, performing the elegant 1957 Meselson–Stahl experiment with Frank Stahl demonstrating the semiconservative replication of DNA. This experiment provided absolute proof of the Watson–Crick hypothesis for the replication of DNA.

SHORE BIRD \ DATE	TALLY	17	18	19	20	21	22	23	24	25	26	27	28	29	30	31	S1	2	3	4	5	6	7	8	9	10	11	12	13	19	16	16	17	18	19	20	21	22	23	24	25	26
PIPING PLOVER	2					2																																					
SEMIPALMATED PLOVER						2	1	2		1			2		8						2	2													5	1					4	1
KILLDEER	4						3	4		6			15		15						3	5													10							
PLOVER, GOLDEN																																										
PLOVER, BLACK BELLIED																																			1							
RUDDY TURNSTONE																																										
WOODCOCK																																										
SNIPE, WILSON																																										
PLOVER UPLAND																																										
SANDPIPER, SPOTTED							2	2		3			2		2						1	1													4							
SANDPIPER, SOLITARY													1		1						2	2			1										2							
YELLOW LEGS, GREATER																					1	1																				
YELLOW LEGS, LESSER							10	10		5			8		10						7	10													5							
KNOT																																										
SANDPIPER, PECTORAL	2						25	20		15			30		40						20	15													10							
SANDPIPER, BAIRD'S																													•					1	2						
SANDPIPER, LEAST							20	15		10			10		10						8	8													15							
SANDPIPER, RED BACKED																																										
DOWITCHER							2	1					1		1																											
SANDPIPER, STILT								1	•																																	
SANDPIPER, SEMIPALM.							50	60		10			90		70						30	25													25							
SANDPIPER, BUFF BR.																																										
SANDPIPER, WH. RUMPED																																										
CURLEW, HUDSONIAN																																										
SANDERLING	4						1						35													5									10	3					8
	M						C.	C. W		N. C.			C. 79	C.							C.	C.				D									C. 79	W					79

Fall 1944 bird list. (Figure continued on following page.)

SHORE BIRD / DATE	Aug 27	28	29	30	31	Sept 1	2	3	4	5	6	7	8	9	10	11	12	13	14	15	16	17	18	19	20	21	22	23	24	25	26	27	28	29	30	Oct 1	2	3	4	5	6	7	8
PLOVER, PIPING															2			2	8	2	2	1	3			1										4							4
PLOVER, SEMIPALM.		6						3										2	10			3			2	6	1									9							20
KILLDEER		6																		25		25		25	25			×															
PLOVER, GOLDEN															26		·	40	2	40		2	25		25	35	4		2														
PLOVER, BL. BELLIED																						1	1			1																	
RUDDY TURNSTONE						1																																					
WOODCOCK																			8																								12
WILSON SNIPE																						1																					
PLOVER, UPLAND														1					1																								
SANDPIPER, SPOTTED																		1	4																	7							
SANDPIPER, SOLITARY		5						4											4		1															2							4
YELLOW LEGS, GREATER																		3		1					1											2							11
YELLOW LEGS, LESSER		5															2				1															2							
KNOT		3																2	4																	3							3
SANDPIPER, PECTORAL																				1																							
SANDPIPER, BAIRD'S																																											
SANDPIPER, LEAST																																											
SANDPIPER, RED BACK																																											
DOWITCHER																																											
SANDPIPER, SILT		5													1	2	1	3	4		1		1																				
SANDPIPER, SEMI-																					1		1																				
SANDPIPER, RUFF BR.																	2																										
SANDPIPER, WH. RUMP.																		3 (2-1)																									
CURLEW HUDSONIAN														25	8	15	30	10	3	25	38		8	25	6	28	3	3															6
SANDERLING		c			E							O	W	79 J.P.	79 J.P.	W	79 C	W	W		79	W	79 J.P.	79	E	79										O ND.							C.

Fall 1944

Graduation day at the U of C in 1947.

Spring Migration of Shore Birds in the Calumet Region 1946

Species	First Date	Maximum		Last Date	Total	Comments
		Date	Number			
Piping Plover	May 4					3 pairs nested at Wolf Lake
Semipalmated Plover	May 4					
Killdeer	Mar 4					S.R.
Golden Plover	Apr 21	Apr 27	8	May 4		
Ruddy Turnstone	May 13					
Woodcock	Mar 13					S.R.
Wilsons Snipe	Mar 25	Apr 27	80	May 4		
Spotted Sandpiper	Apr 27					S.R.
Solitary Sandpiper	Apr 27			May 18		
Willet	May 18			May 19		
Greater Yellow Legs	Apr 13	Apr 21	50	May 12		
Lesser Yellow Legs	Apr 13	Apr 27	50	May 19		
Pectoral Sandpiper	Apr 13	Apr 20	400	May 12		
Least Sandpiper	Apr 21	May 25	25	June 2		
Red-backed Sandpiper	May 4	May 19	65	June 2		
Stilt Sandpiper	May 19					
Semipalmated Sandpiper	May 12	May 25	70	June 2		
Sanderling						
Wilsons Phalarope	May 12					

Spring Migration of Shore Birds in the Calumet region 1946.

Jim and sister Betty in June 1947.

graduation ceremonies in Rockefeller Chapel, Dad and I stayed there for several days. His annual vacations by then had been extended to three weeks. In this way we could be in the park just after dawn when the bird songs were at their best. Wood Thrushes sang in the trees above our lodgings with the plaintive calls of ovenbirds coming from near the moist ground on which they nested.

To our family's great disappointment, Robert Maynard Hutchins, by then the U of C's Chancellor, did not preside over my B.S. graduation ceremony. Two months earlier his marriage had collapsed, and he was temporarily on leave of absence. Dad would have much liked his once-close friend to share in our joy that I was graduating

SLOANE PHYSICS LABORATORY
YALE UNIVERSITY
NEW HAVEN, CONNECTICUT

Sept. 1

Dear Jim—

I have been intending for a long time to write you to congratulate you on the success you made of your undergraduate career at Chicago and to wish you well in your graduate work at Indiana. I am proud of the high stand you have made in college. You seem to have the ability to concentrate on a subject and master it. And you have picked a good specialty. Genetics should be a booming subject for some time because of the possibility of using radio-tracer techniques and the enormously greater dosages of radiations of all kinds from piles and high-voltage machines.

Prof. Muller should be a good man to study under. You should probably stay right there until you have milked his department of what it has to offer. Anyway you cannot plan those things much ahead of time. One usually takes the best offer that comes along when you need it. Just one piece of advice: Drive ahead to a Ph.D. degree without stopping. You should be able to pick up assistantships and/or fellowships to support yourself, and it has been my observation that it is better to make the needed sacrifices while you are young. For a career in science these days you must have a doctor's degree, therefore go directly for it, although there is no

Letter from William Watson to Jim congratulating him on his graduation and noting that "Genetics should be a booming subject for some time…" and advising him to "Drive ahead to a Ph.D. degree without stopping."

[22] After his divorce from Maude Phelps McVeigh in 1948, Hutchins married Vesta Sutton Orlick, his assistant at *Encyclopedia Britannica*. Ever idealistic, he spent the remainder of his career after the Ford Foundation as President of the Center for the Study of Democratic Institutions in Santa Barbara, California. As he wrote in his influential book, *The Higher Learning in America* (1936), he continued to believe that, "If education is rightly understood, it will be understood as the cultivation of the intellect. The cultivation of the intellect is the same good for all men in all societies."

Father and son with the new family Ford in 1949.

Phi Beta Kappa and about to go for a Ph.D. in Genetics at Indiana University (IU). When the year before I received the Ph.B. upon my completion of the Great Books–centered college curriculum, Hutchins instantly recognized Dad despite the more than 30 years that had passed since their freshman year together at Oberlin. For only several more years would he preside over the U of C's fortunes, resigning in 1951 to become Associate Director of the newly formed Ford Foundation. Hutchins saw its large endowment as providing the means to move his educational reforms over all of the United States. In retrospect, Hutchins' years at the U of C were his finest. Never again would he be so obviously the most charismatic person at any occasion in which he partook.[22]

Dad and Mother at last again became car owners in the fall of 1949 through purchasing a new Ford sedan in which I came back with them to Chicago after I finished my Ph.D. requirements at Indiana University. My thesis there was on bacterial viruses (bacteriophages, or phages) in the laboratory of the Italian-born M.D. microbiologist Salvador Luria. During my three academic years at IU, I had the time to bird seriously only during the spring of 1949 when I was a teaching assistant in its Ornithology course. By far the most exciting birding experiences of mine while at IU were sightings of the crow-sized Pileated Woodpecker that long ago had vanished from the Chicago region. On our way to Chicago, we first detoured south to French Lick to let us look in its adjacent state park at the Cerulean Warbler only seldom seen around Chicago. I remained in Chicago for less than a week before flying out to Los Angeles to spend the first half of the summer at Caltech. Then I finished the summer with the phage world at Cold Spring Harbor on Long Island. Dad and Mother drove their Ford East in late August (of 1950) to let the three of us visit Uncle Bill and Betty in New Haven, Connecticut. Afterward we all went to Cape Cod, where Dad and I birded along its Atlantic-facing beaches.

James and Jean Watson in May of 1950 in front of the Luella Avenue house.

Mother and Dad subsequently spent several days in New York City after seeing me in mid-September board the Swedish ocean liner M.S. *Stockholm* that was to take me to Copenhagen. There I planned to spend two postdoctoral years doing research on DNA in the laboratory of biochemist Herman Kalckar. Once my parents got back to 7922 Luella Avenue, they saw all too well that it made no sense to live there much longer.

Watson family birding on Cape Cod in late August 1950—James, Sr., Betty and Bill Watson, Jean, and James, Jr.

Brothers William and James Watson.

Betty and Jim in Copenhagen in the summer of 1951.

Neither Betty nor I would likely ever again have reason to study at the U of C, while Nana, soon to be 90, was increasingly unable to take care of herself and soon would have to enter a nursing home.

Betty's last days on Luella Avenue were in early May 1951, she joining me in Italy when I was temporarily in Naples. She was then emotionally adrift, following the unexpected loss of her Political Science graduate student boyfriend Bob Myers, who abruptly took off for a never-well-defined government position in the Far East. Whether she would ever see him again was not obvious, and she needed a new focus to her life. After we spent two weeks touring Italy and Switzerland, she came back with me to Copenhagen, where she soon found a well-paying job with the American Military Assistance Program attached to the American Embassy. In mid-July Uncle Dudley and his wife, Laura, briefly in

Betty in Paris, May 1951.

Denmark, took us to a dinner at the "Hamlet" Kronberg Castle north of Copenhagen. There he was displaying European culture to some 15 even older Chicagoans that he was leading for six weeks around Europe. Betty found herself again alone in late September when I moved to England to search for the DNA structure at the famed Cavendish Laboratory of Cambridge University. There a small group of physicists and chemists

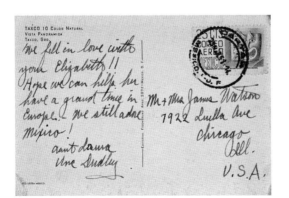

Postcards from Uncle Dudley Crafts Watson referring to Betty's European trip.

Jean and James Watson at Indiana Dunes State Park in April of 1951. Established in 1925, this park has three miles of beach along the southern shore of Lake Michigan and 200-foot-high sand dunes.

were using X-rays to study the three-dimensional structure of biological macromolecules. To my great fortune, its newest member, the 35-year-old physicist Francis Crick, needed no persuading that DNA might be the essence of life.

Once Betty began to be semisolidly entrenched in Denmark, Mother and Dad saw themselves either moving to Hyde Park, where Mother now had a full-time position at the U of C interviewing prospective students for admission, or living near the Indiana Dunes and using the South Shore and South Bend trains to get to its stations in Hyde Park and the Loop. The decision was obvious once they found an unpretentious three-bedroom, wooden house for sale less than a quarter mile from the Tremont station of the South Shore. Less than a 10-minute walk would bring them from its doors to the train that would daily take them back from their jobs in Chicago.

They had no difficulty in soon selling their 7922 Luella Avenue bungalow—South Shore then showing no signs of its subsequent later changing over to all black inhabitants. They were still just living on Luella Avenue when Mother came to Europe for a month in mid-June 1952. After less than a day in London, she found the common hallway bathrooms of the Regent Palace Hotel at Piccadilly Circus not what she had anticipated. So I immediately moved her to the much posher Grosvenor Hotel off Hyde Park. With the American dollar then very strong, my fellowship monies let me not worry about the marked difference in room rates. After much London sightseeing, Mother and I went to Cambridge, where she and I dined with Francis Crick, his wife Odile, and John Kendrew, who thrilled my mother with the inherent quiet beauty of his college Peterhouse.[23]

Upon her return, Dad was already domiciled in their Tremont home with Nana temporarily occupying a ground-floor bedroom. In the fall her

[23] **Writing home from Cambridge, Jean Watson provides a description of her son: "He wears a tie always and his manners are impeccable. His hair is so long it actually frightened me—he has affected a sort of 'Einstein' cut—long and curly—but I managed to withhold comment and am hoping that a word from Max on his arrival in Paris will correct the situation—for it appears a trifle grotesque."**

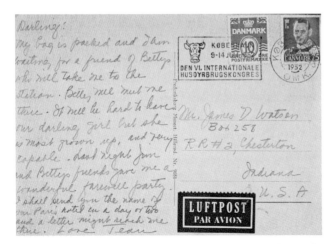

Postcard from Jean Watson sent from Frederiksborg as she was leaving for Paris. "It will be hard to leave our darling girl [Betty] but she is most grown up and very capable. Last night Jim and Betty's friends gave me a wonderful farewell party."

James Watson, Sr., at home in Chesterton, Indiana.

The Watson home in Chesterton, Indiana.

steadily incapacitating, personality-changing dementia (Alzheimer's) led to her being moved into a nursing home 20 miles away in Hammond near the South Shore Line. At last Mother and Dad were living as they long wanted, off a rural road bounded by farm fields and woods, with the small-town shops of Chesterton, Indiana, only six miles inland. Mother soon made herself part of the local community by joining its Democratic Party, while Dad no longer had to go any further than the front porch to hear summer thrushes' melodious songs.

CHAPTER 6

Liberals at Play (1953–1957)

JDW, Sr., and Jean Watson.

Once he began twice-daily, uninterrupted, hour-long commutes to and from the Loop on the "South Shore," Dad had even more hours for reading books and magazines. Always close to his heart were the seminal ideas at the heart of the liberal education experience.

Soon Dad and Mother formed friendships with like-minded neighbors who also took pleasure in ideas and called themselves "polymathics." Roughly once a month, they came together for after-dinner talks in the Sunset Hill home of Colonel Robert Heffron Murray, nearby Valparaiso's most public-spirited citizen, who was born in Chicago in 1881 and educated as a chemist at the University of Chicago. Following graduation in 1904 or 1905, he soon joined the Paramount Knitting Company in Waupun, Wisconsin. Successful in the knitting business, he moved in 1910 to a managerial position at the growing industrial giant Chicago Bridge & Iron Company, having five years earlier, in 1905, married Sue, the daughter of its founder Horace E. Horton. Already a captain in the Army Reserves, Robert Murray was promoted to Major upon his departure to France, and upon returning from active duty once again to Lieutenant Colonel and his later-life designation as Colonel. He returned to the knitting business as founder of the Neumode Hosiery Company, whose presidency he held until his death in 1972. Upon his death at age 91, his home and acres of surrounding land were given to a land trust to preserve its unique ecological features just inland from the Indiana Dunes National Seashore running between Gary and Michigan City.

Dad presented his philosophy of life to his fellow "polymathics" at Sunset Hill on February 22, 1957, preparing for it a 20-page typed

[1] "Your father is engrossed in writing a paper to read before our local 'culture club', its subject, the liberal tradition in philosophy, government and politics. He is finding writing much harder than reading, but really is, I think, enjoying the effort." —Jean Watson to JDW, Jr., February 1957

[2] In addition to the writings of historian Arnold Toynbee, polymath Bertrand Russell, Justice Oliver Wendell Holmes, and philosopher Morris Raphael Cohen, among others, JDW, Sr., was especially influenced by two essays published by the historian Arthur Schlesinger in 1956—"Liberalism in America: A Note for Europeans," originally published in *Perspectives USA* (a Ford Foundation project of Robert Hutchins), and "Conservative vs. Liberal—A Debate," which appeared in *The New York Times* (March 4, 1956), and by a July 9, 1950 *New York Times Magazine* article, "Heresy, Yes—But Conspiracy, No," by philosopher Sidney Hook.

[3] "Your father is putting the finishing touches on his paper which he will read at our meeting Friday night. He has done a very fine job and he had a real feeling of accomplishment." —Letter from Jean Watson to JDW, Jr., 1957

Colonel Murray's Sunset Hill Farm, which he started developing in 1934, eventually totaled 238 acres. During Colonel Murray's lifetime, it was a working farm producing milk, chickens, and eggs that were sold locally. Now a county park, its prairie land and wooded areas are the site of festivals as well as nature-oriented events and hiking.

manuscript, "The Tradition of Liberalism." He started by saying that his long-held liberal philosophy was now increasingly on the defense with its enemies arguing that its application was at the root of much of present-day evils. Dad felt strongly the opposite, believing that the conservative approach to life was inherently untenable in a changing world. In preparing his manuscript, Dad made extensive use of his file of ideas that he had clipped from newspapers and journals of opinion starting from the early 1930s.[1,2,3] Then the onslaught of the Great Depression made it fashionable for many thinkers to question how our world should deal with the great inequalities in the lives of its citizens. Among the thickest of his file folders were those on Liberalism and liberal education. They were even thicker than the clippings about his favorite writer, the lifelong eccentric but highly talented Welshman, John Cowper Powys, whose many novels already occupied a full shelf in the three-door glass, locked bookcase when we were still living on Luella Avenue.

The Tradition of Liberalism, by James D. Watson, Sr.

I would like to talk with you this evening in regard to the philosophy of liberalism. Judging from the criticism directed against this philosophy in recent years, it seems to be defensive. Without exaggeration all of the evils of the present-day world have been placed on its doorstep. And as a result there seems to be a failure of faith in this old traditional philosophy in some quarters.

To those of us who have been liberals for many years it is discouraging to see the ideas with which we have lived discredited. I do not believe that the liberal ideas are the cause of our troubles, or that this philosophy has outlived its day. To me there seems to be enough human reason left to which to appeal against reckless fanaticism of every sort.

What does liberalism mean? What is a liberal? Liberalism is a many sided and troublesome word, and men have used it to hide their views as well as to define them. One of Ibsen's characters remarks: "He has neither character nor convictions nor social position. So liberalism is the most natural position in the world for him." On the other hand, the popularity of this label only a few years ago was evidenced by the replies given to *The New York Times Magazine*. The interviews were undertaken with the object of establishing the meaning of liberalism. The conclusions were indefinite, as to be expected, but one fact at least was established; all those interviewed, and they included Truman, Taft, Dewey and Henry Wallace, said they were liberals.

Before defining liberalism as I see it, and before outlining the problems facing the liberal in 1957, perhaps it would be well for us to go back and set down the historical background of liberal philosophy.

The use of the word liberalism to mean a particular social philosophy does not appear to occur earlier than the beginning of the nineteenth century. But the thing to which the word is applied is much older. The doctrine represents the end of a development which goes back at least to the words of the Hebrew prophets, the teachings of the Greek philosophers, and the ethics of the Sermon on the Mount.

If liberalism means that freedom is essential for the full development of man, then it may be said that there has been a liberal civilization in parts of Western Europe since the Renaissance, in China in the age of Confucius, in Egypt in the Alexandrian period, in India in the period in the rise of Buddhism, and above all, and most brilliantly, in Greece during the age of Pericles. Some of the spirit of liberalism, especially as to the importance of the free play of intelligence, may be found expressed in the funeral oration attributed to Pericles, in the writings of Greek philosophers other than Plato, and in the great naturalistic poem, "De Rerum Natura," of Lucretius.

It is interesting to note that the liberal theory of politics ties up closely with trade and commerce. The reasons for the connection of commerce with liberalism are obvious. Trade brings men into contact with customs different from their own, and in so doing destroys the dogmatism of the untraveled. The relation of buyer and seller is one of negotiation between two parties who are both free; it is the most profitable when the buyer and seller are able to understand the point of view of the other party. There is, of course, imperialistic commerce, where men are forced to buy at the point of the sword; but this is not the kind that generates liberal philosophies, which have flourished best in trading cities that have wealth without much military strength.

Liberalism as we know it today began in the trading cities of northern Italy and in England and Holland, undoubtedly because these cities and these two countries had not been conquered by Spain. It had certain well-marked characteristics. It stood for religious toleration; it was Protestant, it regarded the wars of religion as futile. It valued commerce, and favoured the rising middle class rather than the monarchy and the aristocracy; it had immense respect for the rights of property. The hereditary principle, though not rejected, was restricted; in particular, the divine right of kings was rejected in favour of the view that every community has a right to choose its own form of government.

There was a belief that all men were born equal, and that their subsequent inequality is a product of circumstances. This led to a great emphasis upon the importance of education as opposed to inherited characteristics. There was a certain bias against government, because governments almost everywhere were in the hands of kings, who seldom respected the needs of merchants.

Early liberalism was optimistic because it represented growing forces which appeared likely to succeed, and to bring by their victory great benefits to mankind. It was opposed to everything medieval because medieval theories had been used to sanction the powers of Church and State, to justify persecution, and to obstruct the rise of science. It wanted an end to politcal and theological strife, in order to liberate energies for the exciting enterprises of commerce and science, such as the East India Company and the Bank of England, the theory of gravitation and the discovery of the circulation of the blood.

Mainly, liberalism was anti-authoritarian, it was individualistic, and its one distinctive aim was the liberation of man from tradition restraints. The practical application of liberalism varies with the forces, institutions and traditions which restrain man. The problems of freedom are quite different in a feudal age from what they are in a technological age. The growth of liberalism is the story of man's striving for freedom and dignity.

The evolution of the liberal idea goes along with the progress of Western society from a status-based, church-dominated culture to an ever more democratically oriented civilization. The striving for freedom helped change the structure of society. This meant that the power criticized by liberals shifted from church to state and from state to private concentrations of economic power. Power in one form or another is always the big problem for liberals.

In the late Middle Ages and during the Renaissance, the prevailing society was organized on the basis of status; the rights and responsibilities of the individual were determined by his place in the stratified system. The emphasis was upon conformity.

Early tendencies toward the liberal spirit are seen in protests against this authority. The medieval system was bound to be challenged successfully from the existing order. The challengers were aided by the growth of cities. From this a new middle class emerged.

This new commercial class was the most active force throughout Europe in the struggle for freedom from the restraints of the medieval order. It was the natural enemy of the medieval political organization, primarily because the feudal state constituted a serious barrier to trade. In the late medieval period, therefore, the new middle classes aligned themselves with the monarchs against the nobility.

Once the claim of the monarchy was successfully established, the middle classes turned their attention to means of controlling the "divine right" of the kings they had brought to power. Freedom was now thought of as a problem of the ruler and the ruled. The ruled, in this case the commercial classes, now demanded a rule of law binding upon king as well as subject. The thinking of the first great English liberal, John Locke, was directed to the problem of liberty as it related to ruler and ruled. Locke believed that governments were instituted to serve men. He preached religious toleration, representative institutions, and the limitation of governmental power by the system of checks and balances. These ideas were later to be instrumental in the forming of our own government.

The ideas of Locke later became the basic characteristics of eighteenth-century thought, or what is known as the period of Enlightenment. Liberalism today bears the stamp of this age. The Enlightenment was a period of optimism and revolution, of naïve faith and debunking. It was above all, a period of emancipation in religion, politics, economics and art.

The unifying factor of the Enlightenment was the belief in natural law. The discoveries of Newton had been interpreted as proof that there was a natural order of things in the universe, that the laws of this order might be discovered by human reason, and that these laws furnished standards for the conduct of governments and men. The implications of this doctrine were many.

First, in philosophy, it suggested that the possibilities of human reason were limitless. If reason could discover the laws of nature, there was nothing it might not do. Man could reform himself, his society and his government. And if he could accomplish all this was he not good and, even more important, was he not perfectible? The optimism of this early liberalism was based on this view. Problems were to be solved by an application of reason, and defects of character were to be removed by education.

Second, in politics the concept of natural law became the doctrine of innate natural rights inherent in each individual. The doctrines of natural rights and individualism were joined to produce the belief that all men had the right to possess that which they acquired by their own labor, to speak and write as they chose, to petition and to form combinations, and to worship or not to worship according to their consciences. These principles are set forth in the Declaration of Independence and the American Bill of Rights; and there is no clearer exponent of these ideas than Thomas Jefferson.

Third, in economics the doctrine of natural law again combined with individualism to become the basis of eighteenth-century laissez-faire. The economists, beginning with Adam Smith, maintained that there were certain simple, universal laws governing the economic realm, which, if left to function undisturbed, would bring order out of chaos and general welfare out of private interests.

These simple laws were as follows: (a) All men were born with the natural desire for trade and barter; (b) human actions were dominated by the profit motive; (c) the profit motive stimulated maximum productivity; and (d) maximum productivity was the greatest social good. Therefore, the pursuit of each individual of his own self-interest, or profit, resulted inevitably in the greatest degree of social welfare. Modern liberalism shows itself to be very much the product of the Enlightenment. Individualism, unrestrained independence, and the individual as a law unto himself, characterized the liberalism for that day.

And now liberalism is faced with another problem when it ran into the impact of industrialism. As we have seen the liberal spirit grew out of an essentially pre-industrial environment and yet almost immediately it had to deal with the consequences of the Industrial Revolution.

The economics of the eighteenth century based itself on a natural harmony of interests. If individuals were left free to pursue their self-interest in an exchange economy, based upon a division of labor, the welfare of the group as a whole would automatically result. The logic of such a creed supported the institution of private property. Private property, however, accompanied by the rush of industrialism, led to the development of free capitalism and the institution of the factory and its evils. The development of absentee ownership further accelerated the difficulties faced by a liberalism geared to a small commercial economy.

To meet these problems, liberalism in the nineteenth century split in a number of directions. The career of John Stuart Mill, the most articulate of the nineteenth-century liberals, summed up the transformation of liberalism under the impact of industrialism, from laissez-faire to radicalism to a near socialism. Caught between theory and fact, diehard liberals at first opposed measures such as child labor regulations as unwarranted interference with economic laws and individual liberties.

The moral and economic dilemmas of this position, however, were soon felt, and a new collectivist approach developed. Living conditions of the poor, as Marx and others pointed out, belied the assumptions of classical economics. Soon the concessions of Mill and others signalized British liberalism's withdrawal from doctrinaire individualism toward compromise with the necessity for state controls. Soon in a series of such compromises, British liberalism broadened its base to include progressivism of all stripes, from individualism to Fabian socialism.

Looking back we can see that the liberal forces accomplished much. The feudal system was destroyed. Capitalism replaced the static society of the Middle Ages. A functionless aristocracy was removed from control. Tyrants were curbed. The middle class was left free to employ its energies in expanding the means of production and increasing the wealth of society. Limiting the sovereign power, liberals helped make constitutional government, with its accompanying civil liberties, a reality. Liberalism, as formulated in the eighteenth and nineteenth centuries, was a powerful instrument for progress.

And then coming closer to our own day another new power came into being. It was represented by concentrations of vast wealth in relatively few hands and was used to influence and control government, destroy competition, and increase the unequal distribution of wealth. Here, then, was a restraint to freedom as threatening to the individual as the power of a seventeenth-century despot, which required a new strategy from those desiring to protect individual liberty. It became clear to many that more than political liberty was needed to achieve freedom for man.

It was in these circumstances that a new generation of liberals began to call upon government to restore the balance in society. The new liberalism came to see that the same forces which had once released the productive energies now restrained them.

Twentieth-century liberalism thus tried to adjust itself to the realities of an industrial civilization. It met with both tragic failure and partial success. The failure was in continental Europe. The efforts made were too late and the forces of economic power too great. Instead of liberalism, socialism, communism and fascism swept the working populations and the middle classes.

In the United States, liberalism did not fail, and remains today a political force. An expanding frontier and the blessings of natural resources were partly responsible. They provided greater freedom for action, delayed the rise of the trade union movement and, in turn, severely handicapped efforts of Marxism to gain a foothold here. Partly responsible, too, were a series of brilliant political leaders who helped reshape liberalism into an instrument for dealing with industrial society.

They included Theodore Roosevelt, who first saw the democratic possibilities in big government and the need for big government to meet big business, Woodrow Wilson, and finally Franklin Roosevelt, who completed the transformation of American liberalism from an anti-statest creed to a philosophy willing to use the state to achieve freedom. This theory of the role of government was previously expressed by Abraham Lincoln when he said, "the legitimate object of government, is to do for a community of people, whatever they need to have done, but can not do, at all, or can not so well do, for themselves, in their separate and individual capacities."

Thus liberalism today stands committed to the qualified use of the power of the state to achieve the values of freedom. Modern liberals recognize that concentration of power, whether in private or public hands, is the enemy of freedom. The liberalism of today does not seek the abolition of the price system, but it does seek the regulation and control of the profit system, so as to bring about modifications to suit the requirements of a changing world.

Now the main article in this attitude is the belief in reason as sufficient to solve our problems. With this emphasis on reason many other convictions follow. In politics the liberal believes in civil liberties, the capacity of plain men to govern themselves through discussion and gradual reform. In religion the liberal holds that dogma should adjust itself to the changing insights of science, that Church and State should be kept apart, and that moral insights may maintain themselves in independence of any theology. In economics the liberal wants free trade up to the point where exploitation begins, and from that point on he wants governmental control in the interest of the many.

Since liberals believe in the power of reason to achieve both moral enlightenment and the conquest of nature, they are optimistic about progress. Their antipathies have been for such people as Kierkegaard, Nietzsche, Hitler and Stalin; and more recently for Toynbee, Niebuhr and Maritain among philosophers. The liberal prophets are people like Condorcet, Mill, Benjamin Franklin, Voltaire, Russell and John Dewey.

What is common to the thinking of all of these men are the principles for which true liberalism has always stood—academic freedom, tolerance of heresies, the idea of human progress, civil liberties, the scientific method of free inquiry, and the life of reason.

In the field of government they decry all activities that come under the label of McCarthyism, the restraint of trade, or the bargaining power of labor; immigration quotas, any act which limits the freedom of the press, the educational system or the individual.

At this point let us briefly examine the philosophy which is generally considered the opposite of liberalism, conservatism. Actually the two philosophies overlap in their better sense, as I shall bring out shortly.

Conservatism today exists on at least three different levels. First there is instinctive conservatism, the kind that [as Shakespeare's Hamlet lamented]

> Makes us rather bear those ills we have
> Than fly to others that we know not of.

Needless to say Americans have an ample allowance of this healthy impulse, and nowhere in the world do people talk so much of their "way of life," and their physical well being.

Secondly, there is economic conservatism, the desire to cling to one's economic privileges. There is also plenty of this type of conservatism in America, sometimes concealed, but occasionally expressed with a brutal candor. This narrow vision is most unfortunate for economic conservatism tends to become simply the reverse—not the opposite—of economic radicalism, otherwise known as Marxism. The basic assumptions of extreme economic conservatism and of Marxism are the same: man is an economic animal, motivated by selfish, materialistic desires; social life is a struggle for advantage and even for survival. It is not an accident that the believers in extreme economic conservatism are more frightened by Marxism than are moderate liberals or moderate conservatives; for the economic conservatives can best understand the economic radicals. It is their own fundamental materials that they see reflected in the dark mirror of Marxism, and it is [with] the economic conservatism that the true liberal quarrels. This brand of conservatism must be distinguished from the conservatism that springs from instinct, or that which emerges from mature thinking about human society and history.

For there is a third type of conservatism that is closely related with instinctive conservatism, but is opposed to the materialistic simplifications of so-called economic conservatism. This third level may be called philosophic conservatism, that is, an attitude toward politics stressing the value of traditions, institutions, habits and authority; doubtful of the natural goodness and rationality of man; and based on the thoughtful observation of human affairs and especially history.

The philosophical conservative is suspicious of change, skeptical of progress, convinced of the terrible power of sin, in favor of human nature in its present state rather than some radical revision of human character upon a utopian recipe, reverent toward the past, mindful of the universe a realm of mystery. The conservative is convinced that the liberal Western world is making its way toward what C.S. Lewis calls "the abolition of man."

The *New Dictionary of American Politics* defines conservatism as a reasoned philosophy, associated with the English writer, Edmund Burke, directed toward the control of the forces of change in such a way as to conserve the best elements of the past by blending them into an organic unity with new elements in an ever-evolving society.

I do not believe that liberal principles are today incapacitated. They simply wait their application on a new level. The fact that much of the light of the world has gone out with the wars and the world upheaval, cannot be charged to liberalism as a spirit, but to weak human beings. Nor can liberalism be held accountable because liberals have become overconfident and too optimistic. The opponents of liberalism do not say the Church is dead because of the frequent lapses of Christians.

In the famous third chapter of his *Four Stages of Greek Religion* Gilbert Murray characterizes the period from 300 B.C. through the first century of the Christian era as marked by a "failure of nerve." This failure of nerve exhibited itself in "a rise of asceticism, of mysticism, in a sense, of pessimism; a loss of self-confidence, of hope in this life, and of faith in normal human efforts; a despair of patient inquiry, a cry for infallible revelation an indifference to the welfare of the state, and a conversion of the soul to God."

There now seem to be many signs pointing to a new failure of nerve in our Western civilization. The revival of the doctrine of the original depravity of human nature by the theologian Niebuhr; prophecies of doom for Western culture by the historian Toynbee; the search for a center of value that transcends human interest (Maritain); the mystical cult of "the leader" (Hitler and Stalin); contempt for all social programs, because of the obvious failure of some of them; violent attacks against secularism by the clergy; a concern with mystery rather than with problems (our artists); and above all, the loss of confidence in the method of scientific inquiry—these are only some of the evidences of the new failure of nerve.

I have given as one of the basic principles of liberalism the life of reason. An example of the prevailing distrust of reason shows itself in these anti-rational modes which prevail among so many people today. Reason, it is said, is fallible. That, of course, is not so. Reason is not fallible, reason is infallible. There is no instance of the failure of rational thought. There is, however, a rueful record of the disastrous failures, ruin and misery due to irrational thoughts, to substitutes for reason. The chief task of rational thought has been to rescue man from the disasters brought about by these substitutes.

The liberal must have unqualified devotion to impartial reason. As a well known American philosopher, Morris Raphael Cohen, writes, "our reason may be a pitiful candlelight in the dark and boundless seas of being, but we have nothing better, and woe to those who willfully try to put it out."

The man of liberal philosophy insists on the primacy of thought: the whole dignity of man, according to Pascal; the very proof of man's existence, according to Descartes. Reason is capable of no error that more reason cannot cure. The liberal does not deny the validity of practical sense, or perhaps of the mystic experience; but both must submit to the rule of clear honest thinking: the Catholic Church does not hold children responsible for their sins until they have reached the age of reason.

Another principle for which liberals have always stood is the scientific method of free inquiry. To liberals the critical spirit of inquiry is inseparable from the love of truth that make men free. The method of free inquiry bring intellectual honesty, freedom from prejudice, the courage to consider new and startling hypotheses and the rejection of fakes of all kinds.

The spirit of science is skepticism, and the liberal believes that the world needs the spirit of skepticism to confront dogmatism. True skepticism is the enemy of ignorance and of superstition. Skepticism is not cynicism but it is the awareness that we may be wrong and that what we call truth may be no more final than the infallible truths of the past for which men were willing to kill and be killed.

Science must maintain, as a basic principle, the conviction that one may be mistaken. Scientists are men who take nothing as a final truth without investigation, and the investigation is undertaken without preconceived ideas of what the results may be. The true scientist is concerned with testing his assumptions, he seeks for negative instances to disprove his theories and welcomes the critical comments of his fellow scientists. There are no infallible truths or absolute certainties in science, no principles that must be maintained against the facts. Science is self-correcting.

The liberal's chief concern with the application of science is in the field of ethical values. The charge is now made that the scientific method has corrupted our values and offers nothing in return. I do not think this is true. I believe if any values have been lost, they have been false values.

The scientific method is the way in which a scientist approaches a problem. He endeavors to settle controversy by referring to experience, rather than consulting authority or tradition. He exhibits a willingness to consider all relevant facts; a critical, questioning attitude towards existing beliefs and assumptions, an attempt to eliminate personal bias ar prejudice.

Individually, scientists have all the human weaknesses. But the body of scientists is trained to avoid every form of persuasion but the fact. The object of all scientific work is the search for truth based upon fact. A scientist who breaks this rule, as Lysenko has done, is ignored.

What values grow from this method? Toleration is one which could almost be equated with liberalism and democracy. The basic processes that the scientist uses in his procedures are identical with those which democracy should and to some extent does use, and which are characteristic of a liberal approach. Get the evidence, no censorship is valid; be free to consider all the evidence, free speech; draw the best possible conclusion after hearing all arguments upon the evidence; take that action which seems to offer the most reasonable solution; re-examine the consequences of the action; and guide further action by the improved knowledge thus resulting. This is the method of science, and there, I would say, is the method of liberal democracy.

The society of scientists is simple because it has a direct purpose: to explore the truth, based upon fact. The only effective world community that now exists is the community of science. In this respect, if in no other, the vision of eighteenth-century liberal philosophers has been achieved. It would almost seem that this is the foremost intellectual achievement of modern liberal society.

May I say a word on one more of the liberal principles, the tolerance of Heresy? The meaning of liberalism is doubly important today because communism uses the freedom of a liberal society in order to destroy that society. It is easier to say what liberalism is not, than what it is. For one thing it is not identification with the traditional belief in inalienable rights.

Why is there a failure of nerve on the part of so many people today? Why has there been such a loss of confidence, not only in our ability to achieve the liberal ideals, but in the very validity of these ideals themselves? Perhaps it would help if I gave a short resume of the points made by three important critics of the liberal philosophy.

One of the most influential is the Roman Catholic philosopher, Jacques Maritain. Maritain traces our troubles to the rejection of "eternal truths" that have been given us in revelation. Maritain feels that the liberal's attempt to arrive at such truths by his own powers has led not only to failure, but to the conviction that there are no such truths to be found, that man is the measure of all things, and that morals are a matter of taste. In other words a philosophy of relativism.

Here the liberal dissents. He rejects as invalid the dilemma on which the authoritarian so commonly rests his case: either authority or chaos. He believes that there is an excluded middle. It does not follow that because reason cannot lay down rules that are eternal and unalterable, it cannot tell the better from the worse, or give us any guidance at all.

Another critic of liberalism is Reinhold Niebuhr, the distinguished Protestant theologian. The great fact about man, Niebuhr believes, is his sinfulness, which reason cannot remove or correct. The normal mood of life must be anxiety, and anxiety that can be lightened only by grace from above.

The liberal surely will have none of this. Man will never become perfect; granted. But he can improve himself indefinitely, and what more can be asked of him? As for his submersion in original sin, this seems to be a confusion; the impulses with which we are endowed are neither good nor bad, but neutral, and can be converted with as much certainty into something noble as into some[thing] base.

The third well-known anti-liberal thinker is the historian Arnold Toynbee. His magnificent *A Study of History* brings together all of the criticism of the contemporary outlook we have examined, and adds a few more of his own. Toynbee believes that he has found a remarkable moral pattern in the march of human affairs; he thinks we have one last chance to save our civilization—provided we set our faces in the opposite direction from the one in which the modern liberal West has been traveling.

Toynbee believes that all civilizations have the same goal. They are attempts to transmute gross human nature into a finer substance; they are efforts to transform Subman into Superman, to convert the cities of this world into the City of God. Civilization is "an audacious attempt to ascend from the level of Primitive Humanity, being the life of a social animal, to the height of some superman kind of being in a Communion of Saints. . . ."

According to Toynbee modern man has had an entirely wrong view of his own history and as a result modern society has been a study in social disintegration. The basic assumptions which the contemporary world has inherited from liberalism are reversed by Toynbee at every key point. His philosophy tells us that the meaning of history is not the gradual extension of man's understanding of nature and himself through the growth of intelligence; nor is the gradual advance of freedom and happiness in this world through law and social action its meaning. To Toynbee the meaning of history is the conversion of men's values away from such worldly objectives, and the changing of men into saints.

The conditions that are favorable to material and intellectual progress are precisely the conditions that lead to civilization's decline. Mr. Toynbee's view of history suggests that a pluralistic society is a broken-down society; that the leadership of great individuals is preferable to the rule of impersonal law; that there are no such things as accidents or chance in history; that the moderate reformer, who tries to change a part of society without engaging the entire society, is the Don Quixote of social action.

There is not time to go into all of Toynbee's ideas, even if I were capable of doing so, but I would like to comment on one of them. He seems to insist that difficulty and trouble promote human progress. In the growth of any civilization, of course, there are bound to be difficulties which may be overcome. Victory over difficulties certainly can cement a society together. But this is a far cry from saying that suffering is the reason for human achievement. Pain may be an unavoidable by-product of achievement; but it is not its cause. From my observation, for the mass of humanity, the simple fact is that suffering brutalizes and corrupts. It does not improve men, it grinds them down.

All of these men, and others, are convinced that the liberal philosophy is itself a major reason for our present troubles. There is little doubt that something has gone wrong. But the liberal does not agree that it is the result of a faulty liberal philosophy. As a liberal I believe that the collapse of some liberal hopes is the consequence of fundamental social dislocations that have overtaken us.

War, the business cycle, the Atomic Bomb, and the disorganized international scene have had an obvious influence. We have inward security but outward insecurity. Men are afraid and when there is fear there is a failure of nerve and a tendency to look for certainty. But there have been other movements which are hardly less important.

There is the role that technology plays in our lives. Technology is certainly the fundamental element in modern society. But the decisions on how to use the result of technology (television, for instance), are being made almost entirely by men whose area of responsibility is narrow, and who have to think about only a very few, selected values. The engineers and industrialists who make decisions concerning technological changes have enormous power to affect the quality and conditions of our lives even though they may not know they have this power and have no interest in exercising it. This does not change the fact that their decisions are often decisions about basic social policy. The traditional liberal means of public consultation and consent, on which the authority for such basic decisions have been supposed to rest, have next to no influence here.

These decisions just seem to happen; and to me this is one reason why so many ordinary men and women have come to feel that they are being manipulated by invisible persons whom they do not know and cannot control.

To the liberal this problem is institutional. Industrial society today is an elaborately interconnected affair and this means that decisions made in certain central places have consequences that flow out farther and wider than the decisions made by the most absolute rulers in the past. Those who make these decisions are frequently anonymous even to themselves.

The Madison Avenue account executive does not think of himself as an educator; the industrialist who believes that men out of work should be willing to travel two thousand miles to find a job does not think of himself as advocating an unsettled home life. They do not know that they are making the kind of social decisions they are making; they cannot see most of the consequences of what they are doing; they are likely to feel as much caught in the drift of events as the rest of us.

There are other changes which have tended to change people's beliefs. The constant moving around of our industrial population makes social ties impermanent. The local church is no longer the center of a community. The family is no longer a permanent arrangement in many cases. These changes mean that the sort of independent social group which once stood between the individual and the abstraction known as "society at large" has been broken down.

The liberal society made up of men who set their own standards, run their own lives, and cooperate as equals in dealing with common problems is threatened by such changes as these in our institutions. It would seem, therefore, that the problem for contemporary liberals is to reconstruct the liberal tradition to make it applicable to an age of technical specialization, bureaucratic power, and mass movements. But however difficult these problems are, it seems clear to me that they are institutional, not psychological—political, nor moral.

I do not think that these and other problems can be dealt with by reaffirming our faith in absolute moral principles, or by reducing our faith in human potentialities, or by admitting that rationality is a myth, or by waiting for spiritual change, or in Toynbee's language, transfiguration. The revival of liberal hopes depends upon their being attached to specific programs and definite objectives; it does not depend on disparaging the technical powers and human intelligence which must be our main instruments in dealing with our problems.

James D. Watson, Sr.
Feb. 22, 1957

CHAPTER 7

Life without Jean (1957–1959)

I was just starting my life as an Assistant Professor of Biology at Harvard, living in a one-room apartment carved out of a large wooden home on Francis Avenue, less than 1,000 feet from my lab and office on nearby Divinity Avenue. Over the past Christmas holiday, I had gone back to England to be with Naomi and Dick Mitchison's family at their Kintyre home on the west coast of Scotland. When going off I felt guilty that I had not chosen to be with Dad and Mother in northern Indiana. But I still had more close friends in England than at Harvard, and being with their similarly aged Mitchison sons, Murdoch and Avrion, and daughter

JDW, Sr.

The Mitchison's Carradale House. Dating from the 18th century, the house was expanded in 1844. Naomi Mitchison in a 1994 letter to Joe Sambrook said, "Jim was happy here. He liked the house full of books and babies, when anything might go if it was for learning or happiness."

[1] "Received wonderful letters regarding Mother.... Jean was certainly a very fine person and we can be proud of her, but I can't get her out of my mind." —Letter from JDW, Sr., to JDW, Jr., 1957

[2] "As you know, I stayed with your parents several times, and gradually got to know your mother. The more I knew her, the more I liked her; also I came to realize the central position she took in the life of your family. You and your father, and Betty, must feel a very great loss." —Letter from Av Mitchison to JDW, Jr., May 25, 1957

Lois would almost compensate for my lack of a suitable girl to share my life with. Moreover, I had every reason to believe that Mother and Dad finally had the life they had long wanted.

Sadly, it was to last for only four more months. On the morning of Wednesday, May 8, 1957, I was still in my apartment when Dad shakily telephoned me to tell me that Mother had died of a sudden heart attack several hours before. That evening I was on the overnight New York Central train to Chicago taking me to Gary, from where Dad drove me back to Tremont. The next day Mother's funeral was held in the local Catholic Church in Chesterton. A local priest read some prayers, but there was no Mass or false senti-ment. This is the way Mother would have wished.[1,2]

Portrait of Jean Watson.

All of her many Michigan City Gleason relatives were there to mourn the loss of their beloved Bonnie Jean, as was a large contingent of her University of Chicago coworkers and friends. All the polymathics were there, including Colonel Robert and Sue Murray, representing close friendships formed during the five years they had been living next to the Sand Dunes. Uncle Stanley's wife, Anna Mary, came out from Chicago to see that Dad and I were fed afterward. She all too well knew what it was like to be on one's own. In the winter of 1948, just five years after her marriage to Stanley, he died from a malignant melanoma for which there was no treatment. She was left caring for their two sons, Stanley Ford, Jr., and Donald, as well as her son Charles from her earlier marriage.

Before flying back to Harvard, I arranged that Dad would come to be with me over the long Memorial Day weekend. Then we would drive up to southern New Hampshire before ending up at Woods Hole on Cape Cod. In the meanwhile, decade-long friends Dwight and Anna Sanders, who lived in nearby Ogden Dunes, assured me that Dad would not be alone too many evenings.[3]

Betty, then living in Indonesia, did not even know of Mother's death until after the funeral. A careless mistake by a U.S. Embassy mail clerk resulted in my cable only reaching her after I was back at Harvard. Now she was very happily married to Robert (Bob) Myers, who all too suddenly had vanished from her life to work for our government in the Far East just after her 1949 graduation. Bob's parents were Indiana-born, his German-ancestry father a successful butcher in Monticello outside of Lafayette. Always a super student, Bob was at DePauw College in southern Indiana before he was drafted into the Army, which sent him to Far Eastern language classes at the U of C before sending him to inner parts of China beyond the reach of the Japanese army.

Betty did not hear from him for more than four years until he came to Copenhagen and asked her to marry him and live in Japan. There he worked out of offices of the American Embassy and in Yokohama. Before saying yes, Betty joined me

Betty in Cambridge in 1953, here with Herbert "Freddie" Gutfreund (right) and brother Jim (left).

in Cambridge for several months, arriving just before the double helix was found and using her typing skills to prepare the final manuscript sent to *Nature* on April 2.[4] By the time it came into print on April 25, 1953,

[3] "… I feel that Dad should move slowly before changing his environment and giving up his work at LaSalle. Although not old, he is no longer young. Finding a job would be difficult and the adjustment to it might prove to be even more difficult…. I certainly agree that Dad must sell the house … it is altogether out of the question for him to live alone out there for any period of time." —Letter from Betty Watson Myers to JDW, Jr., June 24, 1957

[4] "The final version [of Watson and Crick's manuscript] was ready to be typed on the last weekend of March. Our Cavendish typist was not on hand, and the brief job was given to my sister. There was no problem persuading her to spend a Saturday afternoon this way, for we told her that she was participating in perhaps the most famous event in biology since Darwin's book." — Watson JD. 1968. *The Double Helix: A Personal Account of the History of DNA*

UNIVERSITY OF CAMBRIDGE   DEPARTMENT OF PHYSICS

TELEPHONE
CAMBRIDGE 55478

CAVENDISH LABORATORY
FREE SCHOOL LANE
CAMBRIDGE

February 22, 1952

Dear Folks

I must apologise for my delay in writing. I have been very busy, largely in the lab, though to a lesser extent I have been going to parties or to peoples homes for coffee and tea. I have also been to London twice in the last ten days, both times for scientific reasons. On Friday I finally sent off a manuscript on bacterial genetics to Delbrück and so I should have have free time. However I becoming quite involved in some new work on the structure of chromosomes and at present suspect I may have solved their structure, though for several weeks I must keep my fingers crossed. I'm in direct competition with the great Pauling who has just published a paper (completely wrong!!) on this subject. Noone has curt beat him to a structure and so it would be very nice to beat him. He comes here in early April and I'm trying to finish the work before he arrives.

Betty seems very pleased with her new rooms, which is by far the nicest I have seen in Cambridge. The train is passing by very rapidly and will be sure in 3 weeks. The

Letter sent to JDW, Sr., and Jean Watson from their son February 22, 1952. "… I am becoming quite involved in some new work on the structure of chromosomes and at present I suspect I may have solved their structure…. I must keep my fingers crossed. I am in direct competition with the great Pauling who has just published a paper (completely wrong!!) on this subject."

A STRUCTURE FOR DEOXYRIBOSE NUCLEIC ACID

Y OK for folm & reprints

We wish to suggest a structure for the salt of deoxyribose nucleic acid,(D.N.A.). This structure has novel features which are of considerable biological interest.

A structure for nucleic acid has already been proposed by Pauling and Corey[1]. They kindly made their manuscript available to us in advance of publication. Their model consists of three intertwined chains, with the phosphates near the fibre axis, and the base on the outside. In our opinion, this structure is unsatisfactory fo two reasons: (1) We believe that the material which gives the X-ray diagrams is the salt, not the free acid. Without the acidic hydrogen atoms it is not clear what forces would hold the structure together, especially as the negatively charged phosphates near the axis will repel each other. (2) Some of the van der Waals distances appear to be too small.

Another three-chain structure has also been suggested by Fraser (in the press). In his model the phosphates are on the outside, and the bases on the inside, linked together by hydrogen bonds. This structure as described is rather ill-defined, and for this reason we shall not comment on it.

We wish to put forward a radically different structure for the salt of deoxyribose nucleic acid. This structure has two helic chains each coiled round the same axis (see diagram). We have made the usual chemical assumptions, namely, that each chain consists of phosphate di-ester groups joining β -D-deoxyribofuranose residues with 3', 5' linkages. The two chains (but not their bases) are related by a dyad perpendicular to the fibre axis. Both chains

-1-

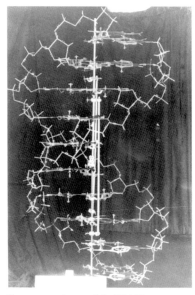

Demonstration model of DNA structure.

Typescript of the first page of the 1953 Nature *Watson and Crick DNA paper. This copy was sent to Catherine Worthingham, Director of Professional Education at the National Foundation for Infantile Paralysis, from which JDW, Jr., had a grant. An earlier version of the paper (around March 21, 1953) had been sent to Linus Pauling. Pauling wrote back saying, "I think it is fine that there are now two proposed structures for nucleic acid, and I am looking forward to finding out what the decision will be as to which is incorrect."*

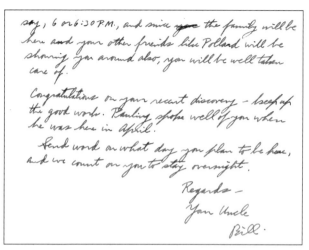

June 3, 1953, letter from JDW, Jr.'s uncle William W. Watson.

[5] "As for the ceremony itself, it was very short; it took only about fifteen minutes, and it was over before we knew it. Bob played his role very well and I believe that we will look back on the ceremony with happy memories. Bob's boss gave me away, and although he was terribly nice about it all, I couldn't help but have the thought while we were walking down the aisle, that it would have been a happier moment if my daddy had been by my side."
—Letter from Betty Watson Myers to "Mother, Dad, and Nana," September 1953

she was already back in Tremont to spend three months with Mother and Dad before flying on to Japan for her marriage in September.[5]

A year later, Betty and Bob's first child, Timothy, was born in Tokyo. Mother and Dad saw their first grandchild when Betty and family came

Betty and husband Bob (center) *on their wedding day.*

back to the States in May 1955. The following year, Bob and Betty lived for a year in Falls Church, Virginia, just outside Washington, DC, until Bob took up a new post in Indonesia. Just before they left again for the Far East, their second child, Holly, was born in June 1956. Knowing he would soon not need it, Bob then sold me his year-old almost sparkling new MG TF two-seater sports car. Before finding a suitable home in its capital city, Jakarta, Betty and Bob lived several hours away in the climatically much more livable, higher-altitude city of Bogor.

Late in July 1957, I took Dad with me for a month to Europe, flying first to Scotland, where I was to be Avrion Mitchison's best man at his marriage on the fabled west coast Island of Skye to Lorna Martin, of Skye's dominant MacLeod lineage.[6] To get there, we rented a car in Glasgow which took us across Rannoch Moor to Inverness and then

[6] Jean Watson had looked forward to this wedding in Scotland in a letter written shortly before her death: "How very nice that Av asked you to be his best man, but the trip does of course complicate it. And a wedding on Skye! I am afraid I could't [*sic*] resist going, even if I had to swim! I still cherish the romantic notion of spending my declining (?) years in the land of the MacKinnons. To this end I shall shortly have an authentic MacKinnon tartan kilt…." — Letter from Jean Watson to JDW, Jr., April 11, 1957

JDW, Sr., with his grandson Timothy in October 1955.

Jean Watson with her granddaughter Holly in September 1956.

Betty in Indonesia in 1957 with daughter Holly. In a letter dated October 2, 1957, she explains, "We moved down to Djakarta about six weeks ago. The house is quite a bit smaller than the one we had in Bogor, but the convenience of it all far outweighs this factor and the added one of the extreme heat."

July 3rd   Carradale

Dear Jim
    If you are really coming over for Av's w dding I do hope we shall see you here. Why don't you come in late July and drive up with me? Or else come back with me? If the former you would overlap Val. It all depends on what your plans are, but I do so hope I shall see you otherwise than at the wedding which will be a crowd.
    I am working like mad trying to get everything done here, and at prese t have no cook so it really is difficult, also I have had a septic thumb and almost the only thing I can still do easily is typing!

                                   Naomi

Letter from Naomi Mitchison to JDW, Jr., planning for the wedding.

The Avrion Mitchison–Lorna Martin wedding party. JDW, Jr., and Av had a long-standing friendship dating from the late 1940s when Av was at Indiana University. Av had written to JDW, Jr., in February saying, "The wedding is to be at the end of July in Skye. It sounds as though it will be rather an enjoyable party, at least for everyone else. The first reaction about wedding arrangements was that there would be no room for anyone, but now barns and outhouses galore are being called to mind. I say this in the hope of persuading you to come over: couldn't you manage to do so? Lorna and I would like it very much if you would be our best man, especially because among my closest friends you are the man that she has known longest." —Letter from Av Mitchison to JDW, Jr., February 19, 1957

Naomi Mitchison at her son Av's wedding. She later related that, "All my family like Jim a lot, but found him something of an innocent abroad, interested in British life but a bit baffled." Naomi belonged to a storied British family deep in science and was the sister of prominent biologist J.B.S. Haldane. A novelist and poet, she wrote more than 90 books. Her marriage to politician Gilbert Richard "Dick" Mitchison produced seven children, of which Avrion, Murdoch, and Denis were biologists. Politics in the Mitchison household were decidedly left-wing. Known as "Nou" to her friends and family, she lived to the grand age of 101.

along "the road to the isles" to the Kyle of Lochalsh. There a short ferry ride took us to Skye, where we were put up in a posh, centrally heated house owned by friends of Lorna's parents. After the wedding in Skye's capital, Portree, we had a week in London, letting us easily go by train for Dad's first look at Cambridge and a meal at Francis and Odile Crick's home, "The Golden Helix." Our last week abroad was in Paris, whose boulevards and narrow streets provided the trip's most enjoyable moments for Dad.

[7] "Yesterday morning I brot [*sic*] all the Books in from the garage. Kind of sad—Mother enjoyed her little room so much. Now that I have my stamps, books and music around me I'll get along but I certainly realize now after 6 months what it really means to be alone." —Letter from JDW, Sr., to JDW, Jr., November 1957

[8] "Depending on when I leave LaSalle and your getting a house, my main enthusiasm is going back to France starting in the very early spring in the south and gradually working my way north to Paris. It does me no harm to let my imagination run wild." —Letter from JDW, Sr., to JDW, Jr., September 1958

[9] "I worked over the weekend getting rid of stuff and preparing for the packing. I am trying to get all my clippings and papers in one 4-drawer file but it seems impossible. The Wilson Bulletins, Auk, Bird Lore, etc. are together." —Letter from JDW, Sr., to JDW, Jr., September 1958

Though Betty initially wanted Dad to join her for a long visit to Indonesia, Dad knew he had to stay at his LaSalle Extension University job until he better sorted out his future options for where to live. Coming back for Christmas 1957 was not a viable financial option for Betty, and I took Dad with me to Nassau in the Bahamas for the week between Christmas and New Year. There we stayed in a small guest house across from the then place to stay in Nassau, its beachside British Colonial Hotel. Unexpectedly much fun to be with at our much-less-expensive lodging were three brash young Yale Medical students, including Max Gottesman, who later chose to practice Molecular Biology as opposed to Medicine at NIH, where he still remains one of its most clever intellects.

A month later I came back to the Midwest to give the Miller lectures of the Bacteriology Department of the University of Illinois. My Ph.D. thesis advisor Salvador Luria had moved there from Indiana just after I completed my thesis requirements in 1950. Once finished with my lectures, I planned to drive up to Tremont to be with Dad for a long weekend.[7] Halfway through my lectures, however, I learned that Dad had come down with bad pneumonia and was in a nearby hospital. Immediately I drove up to see him, fearing correctly that he was also having a nervous collapse brought about by his being too much alone.

Through the help of Dwight and Anna Sanders, a live-in aide stayed with Dad until Betty came back with Timothy and Holly from Indonesia in early May.[8,9,10] Bob was to return four months later, having obtained a year's leave of absence from his then-never-openly identified foreign service position to complete his Ph.D. degree in International Relations at the U of C. With Betty about, Dad again began commuting into Chicago for his job with LaSalle. By summer's end, the Tremont home was sold, Betty renting an apartment near the U of C for her, Bob, and their

Salvador "Salva" Luria (1912–1991) at the 1951 Cold Spring Harbor Symposium. A pioneering geneticist, Luria was driven from Europe by Nazi anti-Semitic policies in 1940. He met up with Max Delbrück soon after arriving in the United States, and they began doing bacteriophage experiments together. He was a founding member of the famous Phage Group that met during summers at Cold Spring Harbor in the 1940s. Jim Watson was his first graduate student at Indiana University in 1947. Luria, Delbrück, and Alfred Hershey won the 1969 Nobel Prize in Physiology or Medicine for their "discoveries concerning the replication mechanism and the genetic structure of viruses." Salva spent his later career setting up the Cancer Center at the Massachusetts Institute of Technology.

[10] "I seem to be back to normal if my interest in ideas and the new books is any indication, came back to the office yesterday with *Harpers*, *The Listener* and the *New Statesman* articles that interested me. We are having good times together, the children come up to my room frequently to talk with Grandfather Watson. I feel very sad that Mother is not here to enjoy them and help Betty." — Letter from JDW, Sr., to JDW, Jr., February 1959

children to live in for the coming year. In turn, Dad moved to a room at the Blackstone Hotel on South Michigan Avenue, just blocks away from his work. For a year more he remained working at LaSalle until he took early retirement at age 61 to move to Cambridge, Massachusetts. There I had arranged for him to live in a tiny, back-from-street house at 10-1/2 Appian Way, only some 1,000 feet from Harvard Square. It was next to my long minimally furnished apartment on the second floor of 10 Appian Way that I first occupied the year before.

Before joining me in Cambridge, Dad joined a classy five-week long "Olson" European tour organized out of Chicago that went first to

The house at 10 Appian Way in Cambridge.

England on the Queen Mary. This, his first organized tour experience, was an unqualified success, giving him all-day to late-night company, some 20 Americans of all ages being his companions. Upon its completion in France, he stayed in Europe for six more weeks, first touring Greece with his physicist brother Bill's younger daughter Alice, who was about to start her junior year abroad with a Smith College group in Paris.

Postcards from JDW, Sr., to JDW, Jr., during his 1959 European tour.

Advertisement for Chicago-based Olson Travel Organization's "select," "deluxe," and "luxury" European tours.

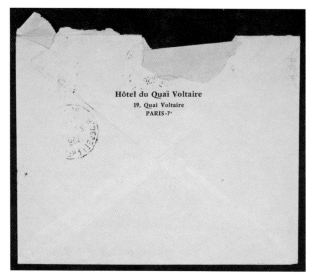

JDW, Sr., wrote frequently to his son during the course of his 1959 European tour. He seemed most happy when he reached his beloved Paris and was ensconced at the Hôtel du Quai Voltaire. He wrote, "Paris needless to say, was very lovely and I entered it late in the evening. The softness of the air, the old fashioned street lights and no neon signs such as fills all of Athens."

Palace Hotel, Scheveningen, Holland—July 1959

The crossing from England was smooth altho we all took pills because someone said the channel was rough. On the way to Scheveningen we stopped on at the Royal Delft ware factory—you know the Delft blue Vermeer used in his paintings. Most of the Dutch use English as a second language and we have no difficulty. We also stopped at the Hague and most of the party (who must see everything) went in the Peace Palace. Thursday was a long day—we were driven to Stratford and Oxford, and I was very tired and have resolved not to go in to every building. One of my best memories of England on this visit will be the great pictures at Warwick Castle—must be worth millions. The food was bad—the gin and tonics were better (but warm).

The members of the group are a curious lot. I try to talk to them all—a judge from Toledo, three house-mothers from Ann Arbor, six or seven school teachers, two Jewish couples from Miami, a rich widow from South Carolina with her daughter, and four college girls who should not be with the group at all. The group leader, an Italian artist from New York City, is competent. He conducts three tours a season and paints the rest of the year.

Grand Hotel et des Iles Borromees, Stresa, Italy—August 6, 1959

Our group leader is a very nice chap—a painter from Greenwich Village and we have become good friends. In the evening we go out with a few of the more lovely members of the group and dance and drink the local wines. I and Tony are the only eligible males and I try my best—it seems to agree with me and I have actually lost a few pounds.

Alice will meet me in Milan tomorrow noon and I'll probably have to take her to Greece early in Sept. Also several of the group are flying to Madrid Aug. 27 and want me to go with them. If I do all this I'll need more $ and I thot Jim could either bring it to me when he comes or send it anyway.

Savoy Hotel, Florence—1959

We arrived from Venice last evening and after drinks with my girlfriends and dinner we walked down to the Bridge of the Shops and then back to the square where we sat and listened to the orchestra until midnight—It was lovely. This morning we were driven to the hills above Florence with the Cyprus trees and the lovely villas then to a very lovely church and the galleries and

"The food was bad—the gin and tonics were better."

"I and Tony are the only eligible males and I try my best."

to a leather factory run by the Franciscans where I made my first important purchase—guess what? This afternoon I slept and what the evening will bring I do not know. I have turned into an escort for the girls—they are afraid to go out alone—these Italian men!

Venice was charming altho I would not want to live there. But I thot my first view of St. Marks Square at night was the most beautiful of the trip thus far. There was an orchestra playing on each side of the square. In the daytime, however, the dirty pigeons spoil it.

Albergo Palazzo & Ambasciatori—August 14, 1959

Yesterday we were driven to Pompeii and then continued to Amalfi for a very fine luncheon at Cappucchini Monastery. Truman and Churchill had lunch there a year ago. We then went to Sorrento via the Amalfi drive and back to the train in Naples and we arrived here at 11:30 last evening. It would all have been very wonderful if the crazy Italian driver had only driven more slowly. The noise and the dirt are here along with the noise.

I have tried to talk and eat with all members of the group and everybody has been very nice to me. I am eating little and it seems as tho [*sic*] I am losing some weight. Some evidence of the 'gup' tummy this morning—I hope it does not get bad—I am sold on the tour especially for older people. Everything to the slightest detail is taken care of and the Olson people certainly think of everything. One of their representatives is waiting for us—wherever we go. The objective to the tour is obvious—herded around, not in one place long enough, and too much time wasted in visiting local industries for shopping. Tony (the leader) says that this trip would cost an individual—if going by himself—2½ times the price. Tony is a wonderful leader, always cheerful altho he must be bored at times.

Windsor—Etoile, Paris—August 25, 1959

Finally I am in my beloved Paris. I seem to feel the same way about it as I did two years ago. It is cooler here altho the leaves show the effect of the very hot summer. Paris is so beautiful in great part, I think, because of the trees—street after street lined with them.

Very good news this morning, a cable from Jim—he is passing thru on 27th and he will see me here at the Windsor. I don't know how long he will be here, maybe only for a few hours but I am all excited.

"Paris is so beautiful in great part, I think, because of the trees—street after street lined with them."

Hotel Windsor Etoile, Paris—1959

Jim arrived Thursday morning by jet and he met some of my new friends—But I had been without sleep. He took a room and left yesterday by train for Geneva. He will be in Copenhagen Sept. 13–18 and he may come back to Paris for a few days.

After I left Jim yesterday morning at the Gare Lyon I walked steadily until 3:30 and ended up at the Place de Concord, where I took a taxi—could not walk it to the hotel. Stopped in Smiths English book store on the Rue rivoli and some English journals. I must get my mind working again. Right now it is a mess.

El Greco Hotel, Athens—September 5, 1959

"In the seats of the world's first theater where the drama was born."

We have just been on a guided tour of the Akropolis and the Parthenon. The party was small and the girl guide gave us a great deal of information. As you know the Parthenon was in very good shape until the seventeenth century when it was bombed by the Turks and not too much of it stands without reconstruction. The original large marble stones lie around the ground and it would be very costly and difficult to bring it back to its original state. We sat in the seats of the world's first theater where the drama was born.

El Greco Hotel, Athens—1959

"In many ways Athens is most like America."

The weather is perfect this evening and the streets are filled with excitable people. Here in this poorer section we walk the streets dodging on to the narrow sidewalks when the cars come by. Mostly American—in fact—in many ways Athens is most like America our goods are everywhere.

Alice was away this week on a four day conducted tour and I have been walking to all parts of the city, and also reading the American Library.

At the lunch table in the hotel this week I have talked to a German insurance man from Weisbaden, a Swedish writer who has studied at Columbia, and a well known British ornithologist who was most pleased to talk to a fellow bird watcher. He is going out to the islands to observe the fall migrations.

Hotel du Quai Voltaire Paris—September 21, 1959

The French are very backward by our standards, very little new buildings but this contributes to the charm. Yesterday Jim and I were looking at prints in a shop off Blvd. St. Michel and when we came out around 5 o'clock we both noticed the beauty of Notre Dame at twilight. They call it the blue hour here.

I have an impression that Europeans mind their own business. No doubt of it you could walk down a Paris boulevard with one leg of your pants purple and the other pink, and people would not gawk at you. As you no doubt have noticed Paris seems to own my heart.

Paris—September 1959

But I am just back from what is surely my favorite museum, the Musee de Jeu de Paume located on one of the corners of the Place de la Concord. This is where the great impressionists have been collected and where sunshine seems to be pouring from the walls. Cezanne, Renoir, Monet, Manet, Degas, and Picasso. Jim was here last Sunday and he told me to be sure and visit it. The use of light by the impressionists is so effective, and after several months of exposure to nothing but religious paintings these were good to behold, my philosophy being what it is. These men seem to announce that life is good, women are delightful to look at, that food is good to eat, and that wine is good to drink. In the pictures here you seem to find a corrective to our times, with their false values.

The pictures are all superb, but two in particular interested me. The huge *Lunch on the Grass* by Manet. The bearded, fully clothed gentlemen seated beneath the trees and the magnificent nude girl, staring out at you. She seems to say "Do you see anything wrong, I don't."

The other was by Degas and it is simply of two laundresses ironing. I noticed the pink, golden later-afternoon light on them. There is a celebration of the sensual everyday life in the picture. As I stepped out on the street, the people seemed dull colored.

Alice is coming tomorrow evening and I'll want to take her and go again to this lovely building.

I have been looking out of the window noticing the speed of cars here in Paris. I have to be alert at all times. There is a cynical joke about this in Paris "To hit a pedestrian in the street—that is sport. But to hit him in the clous— that is sadism." The clous is the crossing between the curbs that is marked by iron buttons for the use of the people on foot, and you have the feeling that thousands of Parisians have met their end in these deceptive sanctuaries. On foot the Frenchman seems courteous, but mounted he is murderous.

"I have an impression that Europeans mind their own business."

"On foot the Frenchman seems courteous, but mounted he is murderous."

"I seem to be very happy on the water."

Paris—October 5, 1959

As I believe I wrote in my last letter I think it best I come home and have booked passage on the Queen Elizabeth Oct. 29. My rent there in Cambridge is going on and it is expensive here. I could go to Italy or Spain for the winter, and would perhaps do so if I had a companion. But alone I don't think it is the thing to do. My French is not strong enuf to carry on an extended conversation, and there is little likelihood I'll make any friends. I realize I am going into that nasty weather in Boston.

I have received some letters from the tour people thanking me for my kindness to them. None of them realized how lonely I was and how good it was for me to be with people. Of all my experiences I like the boat trips the best. I seem to be very happy on the water. I'll have to look into the freighter business.

Paris remains the most beautiful city I have ever seen, and that is also the opinion of everyone to whom I put the question. I'll certainly regret leaving. The other evening I did manage to walk along the Seine to the Place de la Concorde. Dodging the cars I managed to get to the middle of it. In the four directions I could see floodlit the Arc de Triomphe, the Madeleine, the Chambre Deputes and the Louvre. The fountains were playing and it was very charming.

Hotel du Quai Voltaire—October 12, 1959

Last evening I took Alice to the Opera and we saw *The Damnation of Faust* by Berlioz. It was a remarkably fine performance in one of the most beautiful buildings I have ever been in, erected in 1875 and recently done all over. Two stages were used together with a moving screen. They did a remarkable job with music I know so well. But I was kind of sad for it was the Berlioz music I played for Jean the night she died. I often think how much she would enjoy looking in the window with me.

CHAPTER 8

New Frontier Morphs into Vietnam
(1959–1965)

Dad started living in 10-1/2 Appian Way in October 1959 when the second Eisenhower term had only a little more than a year to end. Though 10-1/2 Appian Way started out as a tiny stable that turned into an equally small, two-story residence, it was large enough to house most of the key possessions from his and Mother's Tremont home. Late in October Dad again had at his fingertips the key books, magazines, and clippings of his past.

Though Dad had an almost Eisenhower-like profile, with the resemblance oft noticed, his political leanings never let him even occasionally warm up to Ike's presidency. Only through positive government intervention did Dad believe our country had a real chance to let all its citizens have a meaningful life. The Americans Dad then most admired were Adlai Stevenson and his foremost supporter Eleanor Roosevelt. Each represented for Dad the highly intelligent mind working for social and economic enhancement of all Americans independent of race or creed.

Christmas brought Dad and me to Betty and Bob's new ranch-style home in Vienna on the Virginia side of the Potomac. There we confirmed our hunches that Bob worked for the CIA, whose new large McLean headquarters building was nearby, just off the George Washington Memorial Parkway. Who would be Democratic candidate for President was increasingly at the center of our holiday conversations. Then both Dad and I wanted Adlai Stevenson to again be the Democratic nominee even though Eisenhower had virtually crushed him in the 1956 race. But Bob did not believe that Adlai Stevenson had the guts needed to ward off the Soviets. Still then not to our liking was the junior senator

JDW, Sr.

[1] **Helen Gahagan Douglas (1900–1980) served in Congress from 1945 to 1951. Born into a privileged background, her early years were spent as an actress and opera singer. The Douglas in her name is from her marriage to actor Melvyn Douglas. As a politician she supported civil rights, internationalism in foreign policy, and civilian control of atomic energy. She famously ran against Richard M. Nixon in 1950 for the Senate—he accused her of being "pink down to her underwear," and she bestowed the sobriquet "Tricky Dick" on him. Nixon won the seat with 59% of the vote, and Douglas returned to the theater, the lecture circuit, and writing.**

from Massachusetts John Fitzgerald Kennedy. At the root of our dislike was his being the son of Joseph Kennedy, whose already large fortune helped him become FDR's Ambassador to England in the late 1930s, just before England went to war against Hitler's Germany. Then Joe Kennedy seemed open to a deal with Hitler. Dad and I feared that he had passed on to his son John the same nastiness.

Dad and I, like Betty, then voiced much antipathy toward Eisenhower's Vice President Richard Nixon, whose rise to political prominence came after he defeated the Liberal Congresswoman from Southern California, Helen Gahagan Douglas, through tarring her as a Communist sympathizer.[1] His belligerent anti-Communism might lead us into military commitments that we had no chance of winning. If the United States took on the defense of the autocratic Ngo Dinh Diem–led South Vietnam, we might fare no better than the French. Their long colonial occupation of Vietnam came to an end after their 1954 defeat by Vietnamese revolutionary forces at Dien Bien Phu led to the partition of Vietnam into the Communist North Vietnam and the Western-backed South Vietnam. Then Eisenhower wisely ignored French appeals for help, seeing little gain to the United States in coming to the aid of an inherently lost cause.

Whether South Vietnam could long survive its ever-increasing, anti-government Viet Cong militia was increasingly being debated. Making Dad and I highly skeptical was our reading the English writer Graham Greene's 1955 novel *The Quiet American*. Its main character was a parochial CIA agent unequal to the inherent subtleties of the Far East, where individual human lives seemingly meant so little. Then Dad and I saw the CIA as forces for ever-increasing American commitment to South Vietnam. Much later, in fact, we learned that the CIA was far from a monolithic body. Bob was not alone in thinking that we should never back nations that could not be controlled by bribes. Then it was more than clear that the Viet Cong could not be bought.

During the winter Dad increasingly cooked for me several times each week as he settled into life about Harvard Square. The upscale market of

The Hotel Continental on Harvard Square in Cambridge, Massachusetts.

Sage's on the corner of Church and Brattle Streets was less than 1,000 feet away, and Dad soon mastered the frying pan, cooking pork and lamb chops and using his primitive oven to bake potatoes, especially the sweet potatoes that he knew I would always anticipate. Pies oft heated and served with vanilla ice cream ended many meals. Our favorite restaurant soon became that in the basement of the Hotel Continental, less than a half-mile away and so not too far to walk even on real winter days with many inches of snow. It gave us the solidly American food that Dad never saw reason to fault.

That winter was without our use of a car. Common sense led to my putting my MG TF into a nearby garage from mid-November to mid-April. More winter freedom came when Dad later purchased a VW Beetle that he could park in back of my MG beside the 10 to 10-l/4 to 10-l/2 Appian wooden house complex owned since early in the century by the Theodore Noon family, whose still living patriarch Theodore W., Sr., long ago taught at Lawrenceville. Living with him and his wife was their son Sam, somehow effectively shell-shocked by his earlier life. Though clearly not all there, Sam saw that our boilers always ran when needed and that the sidewalks were never long covered by snow.

Already Dad was making plans for another extensive trip to Europe beginning in late May and lasting through the end of the summer. More than a month would be spent with his brother Bill and wife Betty and daughter Alice, who soon would then be at the end of her postgraduate year studying Art History in Paris. Those winter months of 1960

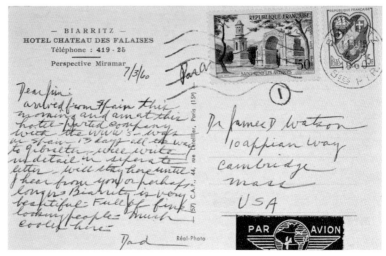

Postcard July 3, 1960, JDW, Sr., to JDW, Jr.

Biarritz, France

*"Arrived from Spain this morning and am at this hotel—parted company with the WWW's—
was in Spain 13 days all the way to Gibraltar.... Biarritz is very beautiful—Full of fine-look-
ing people." "My hotel is on the end and looks down on the old castle. We had a bad storm
Friday and the water at high tide was over the lower drive—very beautiful sight."*

El Hostal Bailén TELÉFONO, 166 N.° 4938

Sr.

Habitación n.° 10

mes de / mois de 6 1960 | 27 | 28

Habitación-Appartement . . 50
« criados-servents . .
Desayunos-Petit dejeuner . . 22
Vino, cafés, licores, etc. . . .
Ropa planchuda-Pressing . .
Conferencias . . . . . .
Garage . . . .

Total del dia-Report 96
Suma anterior-Report . . . 96
Total 128
15 % servicio . . . . . .
Pólizas . . . .
TOTAL. . . .

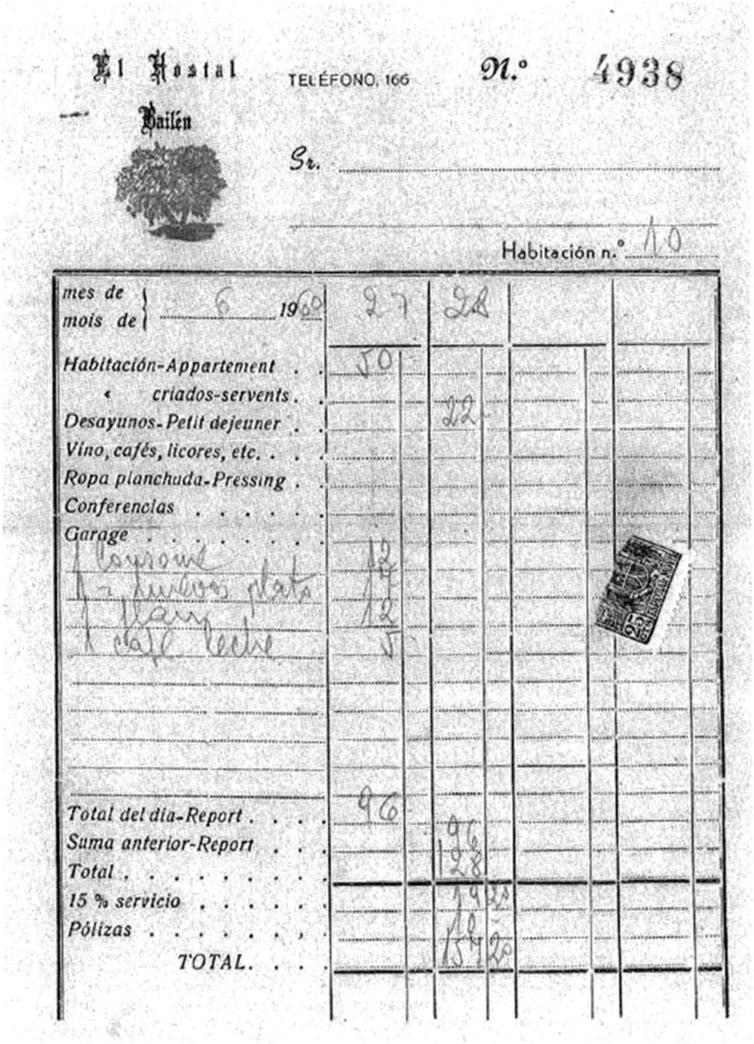

Hotel Bailen invoice

Spain, June 29, 1960—

"Magnificent drive to Gibraltar. Went to the top with the view to Africa and saw the apes. Saw dozen brown sparrow hawks. Had lunch at hotel. On to Cadiz and Jerez where the sherry wine comes from. 443 pesetas including wine and petit dejeuner. Very hot. In Spain you are sent on your way with the old benedictory fare well. 'Quedar con Dios.' "
—JDW, Sr., diary entry

Hotel Chateau des Falaises – Biarritz invoice

France, July 8, 1960

"Storms on the Atlantic. Waves very high. Beaches are closed. Temperature 65. Some rain, wore topcoat. People lined up along the promenade watching the sea. The attractiveness of Biarritz has been recognized for many years. Victor Hugo wrote in 1843: 'I know no spot more charming with its labyrinth of rocks, chambers, arrades, grottoes and caverns. I have but one fear; that it may become fashionable.'"

Biarritz, France, July 7, 1960—

"Walked past the Hotel Palais where Royalty have booked in the past. Edward due any day. This is the most delightful resort. Good hotel, splendid sand beaches, superior food. Very nice middle-class people, no American tourists, the ocean luxury shops, everything. It has been said that one may be seen on the Riviera in the early summer, Deauville in mid-summer but from mid-August to mid-September one must be seen in Biarritz."

HOTEL
CHATEAU des FALAISES
Perspective Miramar · BIARRITZ

SPAIN AND ITS PEOPLE (1960)

James D. Watson, Sr.

Andalusia is what for a century or more the foreigner has understood to be Spain—it is the Spain of the romantic legend. When rolling south to Granada, the road runs through the heart of the country and the distance is about 250 miles. Some of the views are breathtaking, for the color of the land is extraordinary. Against a sky pale blue and grey or brilliant blue, helped with tumbled clouds, brown and purple mountains, with patches of raw earth red as Georgia.

Outside the villages we saw the oxen going round and round thrashing out the corn with their feet, a picturesque Biblical sight—They are blindfolded.

The *paseo* is the sacred evening promenade of Spanish life where people gather by custom to talk and walk endlessly round and round—Spaniards have little social life in their houses. The street is then the place of entertainment—at this hour the women appear and display themselves. External display is very important to the Spaniards—any Spanish crowd is the best dressed in Europe. We see in our mind's eye the Cordoba hat, the woman with the high combs, the proud earrings, the rose or carnation in their hair. We see the gypsy dancer, the narrowhipped bullfighter, the figure of Don Juan. The general standard look is high; indeed, a certain regularity of feature boldness of nose and brilliance of eye—The amount of Jewish blood is, one would think, high (Jewish population, at one time, was enormous in Spain). We see the cool patios of Cordoba, Seville and Granada—hear the guitar and the castanets. We are in the heart of the Moorish kingdom and have one foot in the East—Flowers, singing, sunlight, black shade, and the rustle of water.

The attraction of Spain to the Northerners is its rejection of modern life, its refusal of all that we call progress—The Spaniards have preserved their individualism, a creature unashamedly himself, whose only notion of social obligation is what old custom dictates. The Spaniards have demonstrated that people can survive as personalities without good government, without a sense of corporate responsibility, without compromise, without tolerance—and that in being themselves, they are willing to pay the appalling social price. They are never unnerved by having to face the worst each day.

The Spanish might reply: "We are not an industrialized society, but look at the sickness of industrial man! We have little social conscience, but look at the self-mutilation of countries that have it. Whom would you sooner have, the commissars or the Jesuits? We have digested all that long ago. Spanish skepticism is inseparable from Spanish faith and tho we have a large population of illiterate serfs, we have no industrial slaves. We present to you a people who have rejected the modern [???] and have preserved freedoms you have lost."

The Spanish are people of excess: excessive in silence and reserve, excessive in speech when they suddenly fly into it. They behave with ease as people who live by customs do, and they give an impression of an aristocratic detachment. This is true of all classes from the rich to the poor—There are no class accents in Spain—only regional variants of speech.

There may be enmity [to foreigners] but no open hostility—one meets open, suspicious antagonism in France and Italy, but not in Spain. Where manly [*sic*] welcome and kindness, simple and generous, are always given to the traveler without desire for reward or wish to exploit—the poorer and simpler the people, the more sincere the welcome. The suspicion common in industrial societies, the rudeness of prosperous people, have not touched the Spaniards.

The Spaniards have the melancholy of people who go thru this monotonous life with nothing on their minds and at the present cost of living nothing in their stomachs. They look as though are thinking of some other world, and possibly about death. No other race in Europe has this consuming preoccupation. "Viva la muerte!" was the slogan of the Falangists in the Civil War.

In Spanish painting and sculpture the theme of death is treated again and again by every artist.

Almost all the women appeared to be in mourning and the older men also. People were resigned to the realities of mother nature and of human nature, and in its simplicity their existence was deeply civilized not by modern conveniences but by moral tradition. "It's the custom," they would explain half apologetically, half proudly to the stranger when any little ceremony or courtesy was mentioned peculiar to the people—If things were not the custom what reason could there be for doing them? By bowing to custom they could preserve their dignity. It is that quality, uncommon in freer civilizations, that attracts us to the Spanish.

[2] **Richard Feynman, star theoretical physicist and GLY (glycine) in the RNA Tie Club, had written to Jim Watson in 1960 that he too was "working on messenger RNA half time," adding that "I would like to try to do experiments with you and the bevy of beautiful lab girls!"**

saw my laboratory group using bacteriophage to make a major discovery—messenger RNA (mRNA).[2] Its existence was predicted the year before through experiments at the Institut Pasteur in Paris. I first heard of them at a meeting in Copenhagen in mid-September. Dad's soon going off to Europe for the summer of 1960 let me offer 10-1/2 Appian to François and Françoise Gros, who were coming for the summer to look

Holograph of notes for JDW, Sr.'s Spain essay.

for messenger RNA. Earlier that year in February and March my graduate student, Bob Risebrough, had found convincing if not compelling evidence for mRNA's existence during bacteriophage T2 replication in the bacteria *E. coli*. Most conveniently the synthesis of *E. coli's* own RNA is halted at the start of phage infection. So Bob could observe the exclusive synthesis of the phage-specified RNA whose sole function was likely the transmission of the genetic information specified by the base pairs of phage T2's DNA to the amino acid sequences of their respective T2 protein products. The first outsider I told of my mind's huge discovery was the pioneer of atomic energy, Leo Szilard, whose main research focus had shifted after World War II ended from physics to molecular biology. In early April I saw him in New York City where he was being successfully cured of his bladder cancer by X-ray irradiation at Memorial Hospital. From his hospital bed Leo voiced skepticism as to the meaning of Bob's data, saying he would only believe it to be messenger RNA if we could also show its existence in *E. coli*. So when the Groses arrived in mid-May, I suggested they work on replicating Bob's T2 results in uninfected *E. coli* cells. Unfortunately, their experiment never came to a clear conclusion—*E. coli* mRNA's highly heterogeneous sizes badly overlapping those of the more abundant RNA (ribonucleic acid) molecules used to construct ribosomes, the molecular factories for the making of all cellular proteins.

Key to much of my lab's intellectual and social camaraderie was the no-nonsense, honest personality of my Swiss friend, the medically trained biochemist, Alfred Tissières. He came to my laboratory in the spring of 1957 after almost a decade in Cambridge as a much-sought-after Research Fellow of Kings College.[3] He and his wife Virginia Wachob, whom he married in Denver in the summer of 1958, lived in an apartment building on Craigie Street, a half-mile north off Brattle Street. Dad and I went there the following fall to watch the three Presidential candidate debates between Richard Nixon and John Kennedy on their large TV screen. By then Dad and I both were strong Kennedy partisans, I beginning to favor him soon after seeing him in the flesh as an overseer at the previous June Commencement. If he were to win, key Harvard professors were likely to be in his government, making the Harvard community feel even more certain of its pivotal role in the postwar world.

[3] Jim Watson maintained a lifelong friendship with Alfred Tissières (1917–2003). After study in Switzerland, Tissières came to Cambridge in 1951 to investigate cytochromes, but then left for a short stint at Caltech. He returned to Cambridge in 1953 to concentrate on ribosomes. Tissières returned to Switzerland in 1964 at the University of Geneva's Laboratory of Genetic Biochemistry and then co-founded the Department of Molecular Biology. His later studies focused on heat shock. He was a well-known mountaineer and alpinist, with notable ascents of the south face of the Täschhorn Mountain and the north ridge of the Dent Blanche and an invitation to the 1951 Swiss Everest expedition.

Alfred Tissières and Jim Watson in front of the Harvard Biolabs.

Even before John Kennedy's Presidency came into existence in late January 1961, Betty and Bob had returned to the Far East, their destination this time being Cambodia, where Bob was its new CIA station chief. Dad visited them during the summer of 1961, going by the Dutch passenger boat MS *Oranje*, which he joined in Miami, going through the Panama Canal, across the Pacific Ocean via Tahiti, New Zealand, and Australia to Singapore.[4] There he was met by Bob and stayed for a night at the still very British Raffles Hotel. After their flying via Bangkok to Cambodia's capital Phnom Penh, Dad stayed in an old-fashioned French colonial hotel to which his five-year-old granddaughter Holly brought flower bouquets several times each week. By then she had a younger sister, Lynn, born three months before in Hong Kong, where the medicine could be trusted. My Pan American *'Round-the-World* flight, which I had first used to fly to Moscow for the 1961 International Congress, let me also join Betty and Bob in Phnom Penh in late August. Together with two other CIA families, we visited the extraordinary ancient 13th century temples at Angkor Wat. After 10 days in Cambodia, I flew on

[4] "My long voyage is nearing the end—next Saturday Bob will meet me in Singapore. At the moment land is visible and we will soon head north to the dock in Perth at 8:00 in the morning. I think I'll take a bus to Perth from Fremantle to look the place over—supposed to be a pretty place with the best climate in Australia. Australia is interesting but it has not history or culture—give me Europe every time. But Singapore and Cambodia will be much different—and exciting." —Letter from JDW, Sr., to JDW, Jr., aboard the MS *Oranje*, June 25, 1961

Postcard from JDW, Sr., to JDW, Jr., on his trip to Cambodia.

with Dad to Hong Kong and then Japan, where in Tokyo we stayed at the Frank Lloyd Wright–designed Imperial Hotel, which had remained intact during the horrendously strong 1923 Kanto earthquake.

Dad only remained at 10-1/2 Appian Way for four months before taking in mid-January the Italian liner *Saturnia* from Boston to Gibraltar on his way to Malaga in Spain, famed for its relatively warm winters. A month later, a second Italian liner took him to Sicily, where for three weeks he stayed in a pension run by an English woman in Palermo. Then Dad moved to a chateau near Limoges, France, whose advertisement in the *Saturday Review of Literature* caught his attention. This proved to be the high point of his extended time away, Dad only returned to 10-1/2 Appian Way in June when I came back from Europe, after spending most of April and May in Cambridge, England. Then I stayed at Churchill College as a Visiting Fellow, learning to work with RNA phages in Sydney Brenner's lab at the brand new Laboratory for Molecular Biology (LMB) on the new Addenbrooke's Hospital site. Back for that summer was Alfred Tissières after a year in Paris at the Institut Pasteur. Virginia, his wife, was spending the summer in Denver with her mother before they moved the following year to LMB, marking time while Alfred's laboratory at the University of Geneva was being finished in anticipation of his assuming a new Professorship of Molecular Biology.

Malaga—January 14, 1962
Hotel Miramar

In Lisbon I took the four hour tour of the city, and 30 miles long the Atlantic Coast—was worth it, a lovely day—Lisbon is beautiful with wide boulevards, white houses, ... tile roofs. Of course there is much poverty which we were not shown. A very interesting cathedral—we arrived in Gibraltar at 9 o'clock in the morning rather than 2 o'clock and I had to wait for the Miramar car. But two of the passengers were going to Malaga and they rode with me so the cost was cut to 1/3. It was a drive of 85 miles long the blu Med and the temperature was over 70°—yesterday it hit 80°—much warmer than last year so they tell me.

"The Spanish men seem to spend much time at the coffee bars nibbling at the crayfish."

Malaga—January 23, 1962
Hotel Miramar

I prefer Malaga to the resorts to the south because it is a good sized city with life and a most interesting port. A Polish cruise ship was here the other day and we went all over it. No questions asked about the size of the *Saturnia*—also many freighters.

The business section is quite large with some nice shops but the merchandise is rather shoddy and not cheap—all the streets are very clean and there are some outdoor cafes as in France. All prices of course are cheaper than in the hotel—8 pesatas for café here, 3 on the street. The Spanish men seem to spend much time at the coffee bars nibbling at the crayfish. They are tasty. Many English and Germans here, some American tours (17 day economy).

February 3, 1962
Queen's Hotel, Gibraltar

The Spanish food I cannot take. It was not very good at the Miramar. But there was a good restaurant on the mountain top overlooking Malaga with a lovely terrace 1000 feet up that was perfect and we went up there most every afternoon. It was terraced all the way down with lovely tropical flowers.

Spain is very, very poor and one feels ashamed to be living so luxuriously by their standards—a sales clerk in Malaga will earn no more than 45 pesatas a day. Police were everywhere—they are trying to stop the smuggling of arms. Baggage and cars thoroly searched. But the Spanish are a charming people and I wish them well. But there is trouble brewing.

"I regret they give all credit to Crick."

February 10, 1962

The *NY Times* article I'll not be able to get—will you bring it along when you come or mail a copy to me. I regret they give all credit to Crick. From what you have told me that is not the way it is at all—what are you going to do about it?

February 23, 1962
Taormina, Sicily

My main objection to Taormina is the constant climbing where one has to stop every few minutes to get your breath. But the views are spectacular, especially of Etna on the few clear days. It is very active and one can see the stream of lava.

Just spent a few minutes looking out the window in the rain and saw four common birds here—the goldfinch, the Sardininan warbler, the Black Redstart and the beautiful little Robin. I am pleased to have my binoculars and *Peterson*.

February 27, 1962
Hotel Adriatic, Nice

It seems good to be back in this civilized country again—Prosperity! New stores and luxury shops everywhere and good food!

April 10, 1962
Limoges, France

On the way back we walked down the street in d'Oradour-sur-Glane, the town that was completely burned by the Germans on June 10, 1944. Where over 400 school children died—a new town has been constructed but the old one has been left exactly as it was with the frames of beds, cars, trolleys, etc. standing. Only two persons escaped—a ghastly sight. The Senators in Paris voted to free the soldiers who did it, and one of them was DeGaulle. The German general escaped to the Argentine. The local people have not forgotten.

This is a magnificent section of France and when it gets warmer we intend to take some trips—should be able to keep busy. In the library in Limoges I have discovered the *NY Times Book Review*, the *London Times Literary Supplement*. I am also observing the migrating birds. We heard the first cuckoo and nightingale day before yesterday.

As the summer merged into the early fall, Dad and I nervously awaited the mid-October 1962 announcement of the new Nobel Prizes. While giving some lectures in Harvard in mid-February, Francis Crick told me in great secrecy that Jacques Monod had been asked by the Karolinska Institute in Stockholm, whose leading professors select the winners of the Medicine or Physiology Prize, to write in behalf of our candidacy. Though told not to reveal his recommendation, upon recently seeing Francis, Jacques could not resist the temptation to tell him that we were being considered for the next prize. In turn, I told Dad, saying only that he in Cambridge, Massachusetts, should know. Immediately following the congratulatory telephone call from the Swedish radio telling me that Maurice Wilkins as well as Francis and I were this year's winners in Biology, I popped over to Dad's home with the good news.[5] Soon I

[5] JDW, Sr., received many letters from friends and relatives, congratulating him on his son's Nobel achievement. His good friend Linton Keith wrote that, "the honor has come to him at an earlier age than to most Nobel prizewinners," and then says that Jean Watson "would surely have experienced a thrill of motherly pride." Uncle Dudley Crafts Watson wrote about "this week's excitement" and asked JDW, Sr., to "give my love and congratulations to that splendid, brilliant son of yours."

called Betty, who with her family had just returned from Cambodia to their home outside Washington, asking her to accompany me and Dad to Stockholm. She would much enjoy seeing Francis again and sharing memories of her typing the famous manuscript by Francis and me that *Nature* published on April 25, 1953. Its accompanying illustration of the double helix had been drawn by Odile Crick.

Dad saw his coming for the Nobel festivities as providing an inexpensive way for him to return to the paying guest chateau near Limoges where he so happily spent most of the spring. He and Betty both left Stockholm on the last day of the formal ceremonies, she to return to

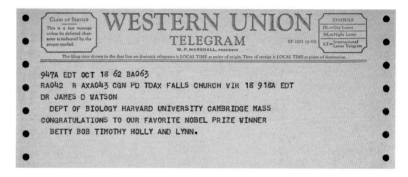

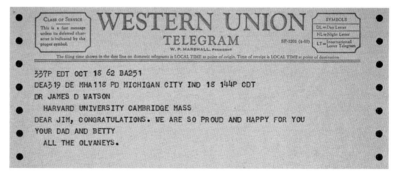

Congratulatory telegrams from sister Betty and family and from Jean Mitchell's relatives, the Olvaneys.

Arrival in Stockholm—JDW, Jr., Betty, and JDW, Sr.

JDW, Jr.'s Nobel medal.

JDW, Sr., JDW, Jr., and Betty speaking with the Press in Stockholm.

Washington and Dad to fly to Paris. After a week in France, Dad flew back to the States to spend Christmas with the Myers while I joined Alfred and Virginia Tissières in the Verbier, whose ski facilities had been built by a consortium put together by Alfred's politician–developer brother Rudolph.[6] On New Year's Day, I flew first to London and then Glasgow to board a DC3 to Campbeltown on Scotland's Mull of Kintyre. Some 20 miles north lay Dick and Naomi Mitchison's home in the tiny fishing village of Carradale.

The ever-increasing U.S. involvement in Vietnam was a bigger factor in my life during much of 1963 than my existence as a most recent winner of the Nobel Prize. My main assignment as a now-more-than-a-year-old member of the limited War Panel of President Kennedy's Scientific Advisory Committee was to learn whether our nation's biological warfare effort might develop effective weapons for incapacitating as opposed to killing our nation's enemies. On one of our meetings in the Executive Office Building next to the White House, I listened to a Green Beret describe the supposed effectiveness of the herbicide Agent Orange in removing the roadside vegetation that made Viet Cong guerilla forces invisible to the American soldiers. Hopefully, the resulting denuded countryside would make the South Vietnamese army recruits less prone to desert once in battle.

By the spring of 1963, I sensed that none of the so-called incapacitating agents then being generated by the biological weapons program at Fort Detrick, Maryland, or by the chemical warfare effort at the Aberdeen Proving Ground would ever be useful tools for maintaining American power in Vietnam. This message I passed on to Bob Myers's colleague John Richardson when he came to dinner at Bob and Betty's home in April 1963 just after his removal as the CIA's Station Chief in Saigon.[7] Hopefully, his removal signified a Kennedy administration decision to stop escalating our commitment to South Vietnam. Though Richardson's career was over, however, other seasoned officials still saw no choice but to win in South Vietnam. Sadly they had the final say. Harvard's former Dean McGeorge Bundy was generally thought to favor hanging tough.

Upon the assassination of John Kennedy in November 1963 and the taking over of power by Lyndon Johnson, Dad and I saw only bad news after bad news coming from Vietnam.[8] What Bob Myers thought was not then in his power to reveal to us. Early in 1963, he had become the second in command of the Far East Division of the CIA under his longtime close associate William Colby, remaining in the position until early 1965. Then he resigned to become the publisher of *Washingtonian Magazine*, which he founded with his University of Chicago roommate Laughlin Phillips, who took on the role of editor.[9] Only many years later did Bob reveal that his resignation followed several years of unsuccessfully arguing that American countryside "pacification" programs would never defeat the Viet Cong. Upon his leaving the CIA, Bob and Betty sold their home across the Potomac, buying a solidly tasteful two-story home on 44th Street bordering Glover Archbold Park.

Dad returned again to France for a month in mid-October 1963, using the French Line for his trans-Atlantic crossings. In Paris he again stayed at the still-then-inexpensive Hôtel du Quai Voltaire, across from the Louvre on the Left Bank, even longer than he lingered at the chateau near Limoges that he found so convivial. Soon following his return to Appian Way, Dad had a sudden, mid-morning, partially paralyzing stroke. Quickly I brought him to Beth Israel Hospital in Boston, where my close friend Howard Hiatt was the Professor of Medicine. Until then Dad thought himself in good health and had not seen a doctor since leaving Chicago. Dad's blood pressure, then above 200, was soon brought under control through medicines that he remained on through the remainder of his life. Fortunately, most of his paralysis soon abated, its only long-term effect being a weakness in the left leg. This necessitated his using a cane when walking, say, to Harvard Square to buy the

[8] Lyndon Johnson got the opportunity to escalate the war after the August 2, 1964 attack by North Vietnam on the USS *Maddox* and the purported attack of August 4th in the Gulf of Tonkin. Congress passed legislation on August 7th giving LBJ the go-ahead to "take all necessary measures."

[9] Laughlin Phillips (1924–2012) was another CIA alumnus who had joined Army Intelligence during World War II and had then worked at the CIA until 1964, with posts in Saigon and Tehran. He was a member of the wealthy Phillips family and, after selling the *Washingtonian Magazine* in 1979, administered the Phillips Collection in Washington, DC, the first American museum focusing on modern art from its inception in 1921. He was director of the museum from 1972 to 1992.

[10] **The Anchorage, Nokomis, Florida: "All of this is going to interfere with the birding but likely he is right and my leg should be better before even thinking of driving—But if I travel again I'll not need one and so you might want to take over the VW and get rid of the MG this spring.... Betty has become very much interested in the water birds and already has a good list." — January 7, 1964, letter from JDW, Sr., to JDW, Jr.**

morning's *The Boston Globe* or *The New York Times*. Upon leaving the hospital, he gradually lost weight and curtailed, though not completely abandoned, smoking.

His Christmas holiday stay with Bob and Betty extended well into January when Betty flew with him to Sarasota, Florida, south of which at Nokomis was "The Anchorage," a modest waterside resort largely frequented by affluent upper-middle-class Midwesterners.[10] There he stayed until early April when he moved back to Washington to be with Bob and Betty for several weeks before returning to 10-l/2 Appian Way. Fortunately his mild weakness in his left leg did not hinder Dad's capacity to drive in his Volkswagen Beetle, using it to stay much of August at a resort hotel on Buzzards Bay at Sippewissett, just north of Woods Hole. By mid-October, he was again back on an Italian cruise liner, which took him to Naples, Palermo, and Lisbon.

Dad's yearly pattern of life became almost routine starting with winters on the West Coast of Florida, summers in Cambridge, Massachusetts and Cape Cod, and the fall seeing him embark on Italian cruise boats to the Mediterranean from which he went to Limoges. In 1965, his European stay ended in Cambridge, England, where I used my 1965–1966 sabbatical–supporting Guggenheim fellowship to almost finish writing *Honest Jim*, the story of how Francis Crick and I found the structure of DNA on the last day of February 1953. Though I had already written its first chapter in the fall of 1962 and the second and third chapters at the end of the following summer, I had put off further writing until I finished writing *The Molecular Biology of the Gene* (*MBOG*). It grew out of lectures for school students that I gave in Sydney, Australia, in the first days of 1964. First conceived as a short 100- to 150-page text, *MBOG* came out in September 1965 almost 500 pages in length.

Only in early November 1965 did I take up residence in an absentee Fellow's suite in Kings College looking out at Clare College. Sydney

JDW, Sr.'s October 1964 crossing on the Italian liner SS Saturnia.

MOLECULAR BIOLOGY
OF THE GENE

JAMES D. WATSON

First edition of **Molecular Biology of the Gene.**

Brenner, by then an influential fellow of Kings, saw that I was given rooms that had good central heat. Then I rented a small car that let me drive Dad around Cambridgeshire after he flew to England from Paris in late November, staying at the Royal Hotel near the Fitzwilliam Museum on Trumpington Street. One evening I took him to Churchill College, where we dined as guests of its prominent humanist and critic George Steiner, then much intrigued by the thought that I was writing a nonfiction novel about the story of the double helix. Upon learning of Dad's extensive collection of novels and literary writings of the Welsh writer John Cowper Powys,[11] George asked him to consider donating them to the Churchill College Library. As a new college there were still many big holes in its collection of important British writers. Immensely pleased that George shared his enthusiasm for Powys, whom many other critics found ponderously obscure, Dad immediately agreed. Within a year he sent his long-cherished Powys collection across the Atlantic.

[11] The prolific British novelist John Cowper Powys (1872–1963), who identified himself as "the English degenerate," is best known for his novel *A Glastonbury Romance*, published in 1932, and his 1934 *Autobiography*. He traveled as a lecturer before beginning his writing career and then lived for a time in New York. He returned to the United Kingdom, settling in Wales and using Welsh mythology in his later works. For some, his writing is challenging and an acquired taste, with his odd twist on Hardy-esque landscapes and strangely vivid imagination. Others admire him as an eccentric modernist writer.

[12] "I hope you have productive conversations with Bragg, Wilkins, and Suzy (a very fine girl). I think you should be able to write the book so these people would be satisfied, there would be no question of libel, and still retain the 'red thread' of life as Barzun put it. Of course, you must be honest and truthful and give a faithful picture of how an important piece of scientific work was actually accomplished and secondly have a book that will be an incentive to a young scientist. These were your two objectives as you told them to me last September. I think you should publish now." —March 9 (no year), The Anchorage, Nokomis, Florida, letter

After the fall (Michaelmas) term ended early in December, I drove Dad down to London where he stayed until flying back to Washington to spend the Christmas holidays with Betty and her family at her home across 44th Street from Grover Alexander Park. Then I went back to Cambridge, hopefully to finish my double helix discovery book before going up to Scotland to spend the holiday at Naomi Mitchison's home. I arrived there, however, with the last two chapters still to put together. The next to last one (28) I wrote in Naomi's study, leaving me only the last chapter to be done at Harvard (to which I briefly returned just after the New Year). I needed only a day to write the first draft of the last chapter (29), which I had decided to end with the phrase "I was 25 and too old to be unusual." The first person to read the complete manuscript was Harvard evolutionary biologist, Ernst Mayr, whom I had by chance met at lunch at the Harvard Faculty Club. Excited by its candor, the next morning he phoned me asking whether he could pass it on to Tom Wilson, the Director of the Harvard University Press, for which he had recently become a syndic (advisor). In turn, Tom also called me within 24 hours saying he thought I had a big winner and wanted my permission to have it read by several more syndics.[12]

I, of course, then wondered how Francis Crick would react to my portraying him so bigger than life. While in Cambridge the previous months, he displayed no interest in the chapters that I had already finished, telling me that he could not help me report exact day-to-day details of how the double helix was found. His past secretary Allison had somehow accidently thrown out all of his correspondence and notes from key years surrounding our discovery. So I thought it best to refrain from sending to Francis its 29 chapters until I could definitely tell him that Harvard University Press intended to put it soon into production and I wanted his comments as to where my memories differed from his.

Knowing that Dad was by then satisfactorily being looked after in Florida, I flew back to London much anticipating soon going for a month or more to East Africa (Kenya, Tanzania, Zanzibar, Uganda, Sudan, and Ethiopia) as a guest of the Ford Foundation. Getting the trip into motion was one of its newest trustees, the Federal Judge Charles Wysanski, with whom (as one of the Harvard 12 Senior Fellows) I dined every Monday night at the Society of Fellows rooms in Dunster House. Very proud of the high-quality staff the Ford Foundation had assembled in Africa, Charlie wanted me to assess its educational programs in action. I was to fly to Nairobi directly in one of BOAC's VC10 first-class seats.[13]

[13] A major battle in 1965 occurred at La Drang in September. In June of 1966, North Vietnamese troops crossed that demilitarized zone and engaged U.S. Marines and the South Vietnamese Army near Dong Ha, but they were pushed back after three weeks of fighting. By the end of 1966, there were 385,000 U.S. troops in Vietnam with 60,000 Navy sailors offshore. U.S. deaths had reached 6,000.

Nice, France, November 1965

On the Promenade des Anglais I was the witness to a lovely action this afternoon. I was on a bench, resting, before continuing to the Quai des Etats-Unis and my pension—a very elderly couple passed and a few feet beyond the old gentleman stopped and kissed his wife on the cheek.

I could not help but think it is love in old age, no longer blind, that is true love. For love's highest intensity does not necessarily mean its highest quality glamour and jealousy are gone and the ardent caress no longer necessary—it is valueless compared to the reassuring touch of a trembling hand.

I see little beauty any more to the embrace of young lovers, and Lord knows I see enough of them from these benches, but the understanding smile of an old wife to her husband, on the kiss I saw this afternoon is surely one of the nicest things in the world—

England, December 8, 1965

Driving down from Cambridge in the early morning (noting the block after block of cheap housing built after the war) we checked into the London Hilton. Needless to say, very plush and expensive. Full of Americans—after dinner with Peter Pauling we went to Covent Garden to see the ballet dance *Giselle*, which they have undoubtedly given hundreds of times.

As is well known some of the finest ballet in the world is given by the Royal Ballet Co. in this historic theatre—It is always jammed to capacity especially on the nights when Fonteyn and Nureyev dance. Tonight Nadia

"Jim, who knows much more of the ballet than I do, was pleased, but the critic in the paper had reservations."

Nerina danced Giselle, and I thoroughly enjoyed her performance. Jim, who knows much more of the ballet than I do, was pleased, but the critic in the paper had reservations.

I read that the queues start forming two days before the box office opens. Covent Garden is surrounded by the warehouses of London's vegetable and flower markets.

Washington, December 27, 1965

There have been a great many people in the Church the past week. How many of them have thought of the great sin of religion in the past—putting into man abject terror and then depriving him of reasonable hopes in this world, insisting that without religion the condition is hopeless.

"I wonder if religion has not done more harm than good."

When I look at poverty and chaos thru out the world. I wonder if religion has not done more harm than good. For one thing, it makes it horrible for the few to exploit the many with ease for the credulous masses are content to look to another world for their reward.

The meek have always been concerned with the treasures that await them in heaven and the strong have been left free to steal the treasures of earth. This process has been going on so long that I wonder if it was not started as a neat stratagem by the crafty few.

Man is the only animal that will tolerate concentration of food or wealth in the hands of the few.

CHAPTER 9

More Than Good Manners (1966–1968)

Knowing that Betty conscientiously went back to her job in the near-by American University Development Office immediately after the holidays had passed, Dad likewise arranged to fly back to Florida to the Anchorage the day after New Year's. On the good side, he looked forward to its food and on most days the temperature went into the 70s. But on the debit side almost all its guests, largely affluent Midwestern businessmen and their wives, were Republicans and intellectual conversations that might make him come alive were nonexistent on the Anchorage scene. So the only recipient he could regularly convey his daily thoughts to was his Diary, expressing them in English—not in the shorthand that he earlier deployed when on many nights he was taking back home still-undone LaSalle work.

JDW, Sr., in Florida.

Florida, January 11, 1966

Acts of kindness from strangers you will likely never see again always make you feel more friendly toward your fellow human beings and are moments of delight.

To mind comes the English businessman who got off the bus and walked several blocks out of his way to guide me to my destination, the boy in Seville who led me to my hotel when I was lost, the man on the *Saturnia* who wrote a letter to the merchant in Gibraltar so that I could cash a personal check, the lady on the *Columbo* who drove me to New York from Boston, the maid in Toledo, Spain, who ran down the stairs with the tip thinking I had overlooked my change, and many others who have assisted me in my travels.

There are thoughtful gestures one does not know about when you start on a journey; they are unexpected awards, but they make travel awarding.

Florida, January 20, 1966

A cold, windy day. Tiring of the same remarks I hear each morning concerning the weather here in Florida and back north, I escaped and sat in front of my cottage, watching the gulls and Terns.

The flight of gulls and Terns is one of the most beautiful things in nature. Their downward fanning sweep is absolute perfection of motions—a movement, and at the same time, a suspension of movement, of which the eye never tires. It is a thing utterly beyond our capabilities, yet it satisfies something deep in us—our response is immediate.

The flight of planes, it is true, provokes our admiration. There is a bit of pride there, too, for they were built by us. But it is not the same thing; we exclaim at the power of the plane, but we are not moved in the same way. The gulls' flight holds the mysterious element, which is poetry.

Florida, February 10, 1966

Why is it so many Americans greatly fear Communism? It seems to me I hear nothing else in my conversations, I read nothing else in the newspapers, I hear nothing else on the radio. Not so over in Europe.

To me it is a confession of lack of faith in our own system and way of life. If we are so superior, if our military might is so great, why all the hatred?

We should be putting our own house in order; we should be proving our case by actual performance and leadership, not in bombing those little people in Vietnam. Has any idea ever been permanently destroyed by war?

"We have made tragic blunders in our foreign policy and undercover dealings."

We have made tragic blunders in our foreign policy and undercover dealings. This is becoming increasingly clear, and we look for a solution to our problems to the very men who made the mistakes.

Florida, February 11, 1966

This morning I had to check the day and the date in the newspaper. I did not know either. But this did not upset me. There is a mild delight merely in escaping from our servitude to Time. A good holiday is spent where the notion of time is vague.

Nowadays we are forced to think in terms of seconds or fractions of seconds and nobody is happy. The society of split seconds is also the society of split minds.

"If you would wander in search of delight, make a start toward a vaguer timekeeping."

So if you would wander in search of delight, make a start toward a vaguer timekeeping.

Florida, February 20, 1966

This evening at twilight I stood at my window watching the afterglow following a magnificent sunset. The loveliest hour of the day. As I stepped out of the door, on my way to dinner, a small flock of Black Skimmers came along, close to the shore.

This spectacular species with the black tipped red bill and the long, narrow wings maneuvers in compact flocks with all the agility and synchronization usually seen only in shorebirds.

It has the curious habit of cutting the surface with the tip of its bill as it flies over the water. By doing this one theory is that it creates a disturbance, bringing to the surface fish which it picks up on the return trip over the same course. They feed only at night or early in the morning and at dusk.

The sight of these unusual birds brings back memories of life ten years ago when Jim and I first saw the skimmer on the southern shore of Long Island.

Florida, February 22, 1966

In answer to Mr. Lilienthal's essay on the population explosion, a lady writes to the *New York Times*:

"Does he know that the birth of a child is life's greatest joy, that watching him grow is its greatest fulfillment? We look forward to many children and we thank God that in this industrialized age of ours where individual anonymity is discouraging, nature has provided me with the means by which we may truly feel worth something to ourselves?"

Granting all of this, I would like to ask some questions! Is personal fulfillment the end of human endeavor?

Is a child a weed, that one just watches him grow, or should one help him by giving him an environment that will give him the opportunity to realize his full potentialities?

Has the industrialization of society caused greater anonymity than crowding?

Does being jammed in like a herring on a subway make one anonymous?

Does it not matter that even with our present population, traffic is jammed, our public schools are overcrowded, our housing is expensive, tasteless and expensive, that city people have to be stacked in ever higher piles of identical boxes?

It is true that a child can give parents a feeling of 'exaltation' but it does not follow that a woman with ten children will feel ten times more exalted than a mother of one.

"It is true that a child can give parents a feeling of 'exaltation' but it does not follow that a woman with ten children will feel ten times more exalted than a mother of one."

"The Egret was in full nuptial plumage. It has a delicate quality, making it unquestionably the most beautiful of our herons."

"It is very noticeable in returning to this country from France or England that our expression, either oral or written, is inferior."

Florida, February 25, 1966

My lack of mobility prevents me from making too many bird observations. But yesterday afternoon both the yellow and night Heron and Snowy Egret were feeding on the oyster bar. They were only 30 feet or so from me making identification easy.

The Egret was in full nuptial plumage. It has a delicate quality, making it unquestionably the most beautiful of our herons. The bright yellow feet are distinctive and I noted that it was constantly stirring the bottom with a quick movement of its foot.

There was delight in watching the yellow Crown. Not a common bird in Florida. Its large white-crowned head and cheek patches are striking, and what a beautiful scientific name—*Nyctanassa violacea*, "Violet Night Queen." This noon as I was talking with my neighbor a flock of Cedar Waxwings flew into the small tree at our side. Then, another flock followed by a third. The branches were bending under their weight. Off they soon went, several hundred of them.

Down here in Florida, many of our common birds are in large flocks. As they go north to their breeding grounds they separate. The Robin, for example.

Florida, March 1, 1966

The other day a lady read to me a letter she had received from her daughter. If I had not known that the girl was a University graduate I would have said it was written by a child in one of the lower grades.

The area in which a poor education shows up first is in expression—oral or written—it makes little difference how many university degrees a man may own. If he cannot use words to move an idea from one point to another, his education is incomplete.

It is very noticeable in returning to this country from France or England that our expression, either oral or written, is inferior. It is seen in the speech of a person in humble positions, one can observe it in the letters written to the magazines or newspapers.

I remember from years ago Madam Thiverin in Limoges stopping our conversation at the lunch table to chastise her twelve-year-old daughter because she had made a mistake in her grammar—at how many tables in America would this have happened?

What are the reasons for this difficulty in our country? One sure is the use of the multiple choice examination rather than the theme. Another may be the conditioning based on speed rather than respect for the creative forces.

A recent examination in a California high school had the student writing four 300 word essays in two hours.

Neatness is valued more than style. Facts are thought more of than clarity and distinction. The slovenly speech heard on the radio & TV do not help. All of this is a long way from the days when Thomas Mann was not ashamed to admit that he would often take a full day to write 500 words, and then spend the next day correcting what he had written.

Florida, March 2, 1966

So many of the guests here at the Anchorage are Episcopalians. [T]hey ask me my faith and I say, "I was raised an Episcopalian, received a prize for seven years perfect attendance at Sunday School." They never seem to ask me if I still am one, and I remain silent. I feel a very real wall between us.

"So many of the guests here at the Anchorage are Episcopalians. I feel a very real wall between us."

Florida, March 6, 1966

I have read recently that the book is likely to disappear, that even the library, in its present form, will be different in the future. This to me, a frequenter of libraries for 56 or more years, and a buyer of books for almost as long a period, is a sad thought.

Whatever its changes, its size, its architecture, the library has remained a magic promise, a mystery of much to be read, an excitement, a continual source of delight. I step into it as I did today, in Venice, Fla; as always with anticipation and eagerness.

It is a quiet place, it costs nothing, one learns a little, and one is surrounded by other people with a like bent. One does not even smoke! What better place could there be to escape from the noisy world for a few hours.

I often think of the small library at Oberlin, the traveling libraries at the front during World War I in France, the main and branch libraries in Chicago where I took the children to get their first cards, the American library in Paris, the libraries in Limoges, Gibraltar and Athens, and now the great libraries at Harvard.

How little of the din of the world enters into a library!

"What better place [the library] could there be to escape from the noisy world for a few hours."

Florida, March 11, 1966

I have received a letter from my son suggesting that I come over to Geneve to join him. Barely back from Europe three months, here I am planning to travel again.

I still find this delightful whereas I very well know that the actual travel itself may be a very dubious enterprise. But with paper, pen, calendar, maps, travel folders and the balance of my bank account in front of me, I still recapture the old bewitchery.

For a day or two I dismiss what I know so well, the indigestion and incredible beds, the service charges and tips, the desperate packing, the cold, heat and rain, and loneliness far from home.

The travel folders say nothing of these things, suggesting only the dawn in the mountains, sunset in the desert, the oranges outside your window, the moon rise over the water, laughter in the night.

Florida, March 14, 1966

It is startling down here in the South, to run up against the race problem. The Boston Red Sox, training over at Winter Haven, had had trouble in housing their Negro ballplayers. In a barber shop up in Sarasota a year ago, not knowing I was a Northerner, I heard a vicious attack against the Reverend King, who had just won the Nobel Prize. My next door neighbor, a faithful churchgoer, never fails to say something derogatory against the 'darkies.' This whole West Coast is not only antiNegro but anti-Semitic.

Florida, March 16, 1966

One should be very careful in pointing out differences between people or drawing general conclusions from insufficient evidence, but I think I am safe in mentioning the following American characteristics.

1. American voices. In England last December a businessman said to me, "How loud Americans speak." I promptly replied, "I am an American and my voice is low-pitched." He instantly said, "You are an exception." In the main I'd say he was right. In the better restaurant in England or France you can almost hear a pin drop, while here in the states it is often a din.
2. The excessive friendliness, especially of Midwesterners. Friendliness, of course, is pleasant at times, when one is in troubles or when one is alone with no companion. But it can also be very irritating when you wish to be alone with your thoughts, or when you feel there is an ulterior motive in the friendliness. It must be annoying to Frenchmen who are 'loners.'

I remember asking the stewardess, on my return plane trip from Japan, if I might have tea rather than coffee and she replied with a wink, "You sure can." I had left the land of courtesy and entered that of familiarity.

"In the better restaurant in England or France you can almost hear a pin drop, while here in the states it is often a din."

Florida, March 21, 1966

I could write of the wonderful weather of the last four days. My sort of weather, temperature in the middle 70s, bright sun, and the breeze from the North. No humidity.

But instead I shall begin to set down the essentials of the simple philosophy I have worked out over these past years. These basic tenets come from many sources in my reading, and I have checked them against the experience of my rather uneventful life.

The following I believe to be true in regard to our social and political living. I have come to believe in the three following formulations: The first is: Never presume that your own meaning and values are the only authentic ones in the world. The second is: to safeguard the uncompromisable, you must compromise the least essential. And the third: when the obtrusion of ideals can only deepen enmity, then you must "contain" them within the privacy of your individual self.

This, to me, seems a very modest code but I know there are those who will call my pluralism indifference, my compromise opportunism and my containment of ideals dishonest suppression. But in doing so they will ignore the finding of anyone who has had to engage in an active life.

And in an age driven further and further by theologians and politicians alike toward a suicidal rigidity (note the uncompromising attitude of our Sec of State), there surely is gain to be found in a position which asks of us magnanimity toward diverse meanings, a moderate amount of ideological generosity, and the private self-reliance which resists all imperialisms.

"There surely is gain to be found in a position which asks of us magnanimity toward diverse meanings."

Florida, March 27, 1966

The memoirs of Field Marshal Keitel are reviewed in *The New York Times*. He was executed by hanging as a common criminal. He lived in an unbelievable world of fantastic automatons all lacking one vital asset. A sense of morality.

"I grew up in a traditional military environment and we were not concerned with right or wrong."

In contrast to all of this, how delightful and pleased I am to read of a position taken on moral grounds. For instance, Martin Luther King's remarks in accepting the Nobel Prize Award several years ago, "I refuse to accept the idea that the 'isness' of man's present nature makes him morally incapable of reaching up for the eternal "oughtness' that forever confronts him."

"I suppose I should write of my fellow guests. I am definitely an outsider as far as they are concerned."

Florida, March 31, 1966

Now that I am in my next to last day here at the Anchorage, I suppose I should write of my fellow guests. This is difficult to do both because I do not know them too well, and because of my own very different interests.

I am definitely an outsider as far as they are concerned. Their values are not mine—not being active, not owning a car, not playing bridge, not fishing or owning a boat, my interests and pleasure are not theirs.

It is in stating at times—the constant din of old people's chatter, the talk at all times of the weather, hearing the same expressions, 'don't get up' as a lady approaches the table, everyone asking 'do you feel better?' If you miss a meal or two: "Did you catch anything?" among the fisherman. The same expressions uttered by Mrs. W, who is senile, the disagreeable rasping voices of some of the women, and the constant probing on their part to find out my beliefs. And perhaps most of all the emphasis they all place on money.

To offset all of this are their perfect manners, and their kindness (if they have accepted you). But none of them read serious books and therefore there can be no serious discussion—what notions they have come from the *Wall St. Journal*, David Lawrence, and the right-wing columnists.

Many of them are extremely wealthy, but they are selfish and their money will go to their children, thus perpetuating another generation or two of the same kind of people.

"What notions they have come from the Wall St. Journal, *David Lawrence, and the right-wing columnists."*

By then Harvard University Press was to be my new manuscript's publisher, and Sir Lawrence Bragg had agreed to write the introduction. Its title was to be *Honest Jim*, opening up comparison with *Lord Jim* by Joseph Conrad and Kingsley Amis's comic masterpiece *Lucky Jim*. Dad was not alone then in wondering whether Francis Crick would go along with its publication. Its opening sentence was to be, "I have never seen Francis Crick in a modest mood." This, however, was the essence of Francis, who scientifically had little to be modest about. No one who had already read the manuscript thought it would in any way harm Francis.[1]

Chapter 2̶7̶6

My scheme nonetheless was torn to shreds by the following noon.
Against me was the awkward chemical fact that I had chosen the wrong
tautomeric form for guanine. Before the unsettling truth came out, I ate
a hurried breakfast at the Whim, then momentarily gone back to Clare to
dash off a note to Max Delbrück. A need to write Delbrück had troubled
me for several days but with the bases on my brain, I procrastinated reply-
ing to his recent terse letter. It reported that my manuscript on bacterial
genetics looked unsound to the Caltech geneticists. Nevertheless the letter
went on to say that he would accede to my request that he send it to the
Proceedings of the National Academy. In this way, I would still be young
when I committed the folly of publishing a silly idea. Then I could sober
up before my career was permanently fixed on a reckless course.

Ordinarily a letter like this from Delbrück would have thrown me into
deep gloom. But now with my morale soaring on the possibility that I had
the self-duplicating structure, I reiterated my faith that I knew what
happened when bacteria mated. Moreover I could not refrain from adding
that I knew what had just devised a novel and biologically interesting
DNA structure which was completely different from Pauling's. Nevertheless
I retained a debt of caution by giving no details of what I was up to.

The letter, however, was not in the post for over an hour before I
suspected my claim was nonsense. I no sooner got to the office and began
explaining my scheme than the American crystallographer Jerry Donohue
protested that the idea would not work. The tautomeric form which I
had copied out of Davidson's book was in Jerry's opinion incorrectly
assigned. My immediate retort that several other texts also pictured
guanine in the enol form cut no ice with Jerry. Happily he let out that
for years organic chemists had been arbitrarily favoring one tautomeric
form over another on only the flimsiest of grounds. In fact, organic

First manuscript page of Chapter 26 of The Double Helix.

[1] In April of 1967, geneticist Guido Pontecorvo wrote to Jim Watson, saying, "It would be a tragedy if the book were not published more or less as it stands. But I can understand Francis', let us say, lack of enthusiasm." He offered to speak with Crick but Jim in a return letter said not to because "I don't want to involve mutual friends in the dispute between Francis and me."

Washington, April 14, 1966

To a baseball game with Bob and Timothy to watch the White Sox beat the mediocre Washington team. How much more satisfactory it is to actually be in a ballpark than watching the game on TV. In my opinion.

TV is an absolute delight to that portion of the audience it values least. These are the very young and the very old, the under-educated and the un-skilled, who have never seen and will never see an unsatisfactory program.

But the educated, the professionals, the enriched, to many of this group of people TV is usually beneath attention, much less consideration.

TV acts as a great leveler, a sort of tomato ketchup on a feast of culture. The great wasteland.

What a pity that this marvelous result of science should be in the hands of a few men whose only criterion is the public poll and rating by number of listeners.

[2] **October 11, 1965, *Michelangelo*, Italian Line: "You are very busy and I trust you leave with all settled to your satisfaction. In regard to the new book remember what I said about calling any living person stupid, etc. You will find it can't be done. I liked your statement of your general purpose to recreate for young scientists the excitement of those months and the way you went about it—can't you record some actual conversations you had with F? But I read the chapters very hastily. I know you will show them to me again." —Letter from JDW, Sr., to JDW, Jr.**

Dad only read my complete manuscript about the finding of the double helix when five months later in mid-May 1966 he joined me in Geneva, Switzerland.[2] There I was finishing my sabbatical year at Alfred Tissières's new laboratory while living in the Hotel des Bergues at the foot of Lake Geneva. The high point of Dad's visit in Switzerland came when he and I joined my former student Ray Gesteland and his wife Harriet for lunch at the famous three-star restaurant Père Bise on Lake Annecy. Though the bill came to more than 40 dollars a person, the American dollar was still very strong, and with my royalties from the *Molecular Biology of the Gene* equaling my Harvard salary, I was spending many less dollars than I made and wanted Dad to enjoy Switzerland at its best.

Geneva, May 15, 1966

There is little doubt the 'American influence' in Europe and elsewhere in the world is resented.

This afternoon in the warm sunshine, I sat at a cafe table in front of the Hotel Bergere. A well-dressed Englishman and his wife were sitting next to me. A group of children passed, many of whom, perhaps most, were wearing the jeans.

He laughingly said, "You Americans are corrupting us. Too much American influence. They are vulgar imitations of your cowboy children." He said of this half jokingly insasmuch as he was the European representative of our American firm, Union Carbide Company.

He made the good point it was an intelligent move on the part of Union Carbide Company to have a European rather than an American to head operations on the Continent to offset the criticism of the 'American influence.'

We also discussed the French reaction when our automobile co. bought the controlling interest of the French company, and the replacement by Ford of five English executives with Detroit men, and the immediate undesirable publicity in the London papers.

Getting back to the children and the blue jeans I mentioned that the mothers put jeans on their children because they are cheap, they wear longer, they are easily washed, they even have a better appearance after they are patched. My friend agreed.

But the fact remains there has been an American takeover. The convenience of the supermarket is replacing the small food shop and the process will likely continue. Western clothing is replacing local styles. The older people are angered and aroused. The younger people like it. It will continue not because it is American but because the replacements are more satisfactory.

"The fact remains there has been an American takeover."

To me all of this is most unfortunate. One does not visit a foreign country to see the identical dress, food and customs of home.

Geneva, May 25, 1966

Bob Dylan made his Paris debut last evening at the Olympia Music Hall. Two thousand ze-ze faces turned out paying 60 francs a seat.

I saw the performance on the TV in the Hotel Bergere. Must say I was impressed. The police were there in force but there was no trouble. Dylan himself saw to that.

"Bob Dylan made his Paris debut last evening."

He took the stage promptly and ignored calls of "Happy Birthday."

He quickly went into his repertory to the accompaniment of his guitar. A huge American flag served as his background.

In his tight-fitting olive suit he looked for all the world like an illustration from a Dickens novel. He took neither bows nor curtain calls, restricting himself to throwing some kisses to the house at the evening's end. Occasionally he played the piano as he rendered his most famous songs.

This strange figure and his haunting 'Desolation Row' number leaves an impression that will linger.

Paris, June 1966

A very pleasant man from London, after a short discussion of the American war in Vietnam, said to me this evening, "It's your war, Watson, good luck." It is difficult for me to recall a single European with whom I have discussed the war, who has not come up with some variant of "you're on your own."

Why is this attitude developing in Europe? One reason is that they see Vietnam as an American battle just as Algeria was French, the Congo, Belgian, and Indonesia, Dutch. Those countries know we did not exactly rush to their aid at that time.

When you bring up the Korean War with their support, they reply, "That was different. It was a UN War, and that was the Cold War. We were directly involved. Times have changed."

The blunt truth is that in 1950 most Europeans felt they needed the US to keep alive; now in 1966, with much less fear of the Soviet Union, they sense no such necessity. They know, of course, that they are resting comfortably under our nuclear umbrella.

Perhaps Television more than any other factor is causing the change in the European's thoughts, especially of the young. In their living rooms and cafes almost every evening, they watch news films of the War.

What they see in the most part, is Americans killing Vietnamese. Their reaction is simply horror and moral repugnance. The crucial impact comes from images on Television. Not words in a responsible newspaper.

The US is suffering from this first televised war. The War is defended by responsible Americans politically, but it is a psychological mistake, especially in Europe. The continuing conflict in Vietnam is working against our interests in Europe.

"They know, of course, that they are resting comfortably under our nuclear umbrella."

Soon after Dad came back for the summer of 1966 to 10-1/2 Appian Way, I took him to the Episcopal Church near Harvard Square for the marriage of my undergraduate tutee Nancy Haven Doe to her long on and occasionally off boyfriend Brook Hopkins. They had begun going together soon after Nancy entered Radcliffe and met Brook, whose major was English literature. Visually Brook and Nancy at first glance were the perfect couple—bright, good-looking, and tall. Nancy's mother Budgy was not so sure the marriage would work, telling Dad at the wedding reception that seeing that we so enjoyed talking to each other

Nancy should have married me. Dad knew from the moment several years before when I first brought Nancy to have dinner with us that we would never be anything more than close friends.

Cambridge, August 1966

It is hard to follow events in this hot summer of 1966 without a feeling of foreboding—there is sickness about.

A murder of insane brutality occurs in Chicago, a berserk student picks off citizens with a rifle in Austin—anger and violence seem just below the surface everywhere.

Is prosperity making American happy? We have the best cars and the worst traffic deaths; the greatest wealth and most frequent crimes—our big cities all have malignant ghettos; our national idealism is coupled with racial riots—we are fighting a miserable war that troubles the conscience.

If you don't believe it, listen to the radio, read the literature, they are filled with the rants of bigots and of charlatans selling quackery.

Everyone, of course, has a streak of violence, would like to bash somebody's face, as things grow more and more complex, it is hard just to be normal—there is hardly a human relationship that has not altered.

It is difficult to know what is expected of us.

We have changed our way of living, but we have not changed ourselves. Our wealth constantly accumulates, always along with it, our mocking frustration.

Nancy Hopkins in 1974. Currently the Amgen, Inc. Professor of Biology at Massachusetts Institute of Technology, with Mark Ptashne she identified the λ repressor operator binding sites and later investigated gene expression in RNA tumor viruses. In the late 1980s she moved to developing large-scale insertional mutagenesis methods in zebrafish. She has been an ardent opponent of gender bias in scientific institutions.

Until the spring of 1963, when Nancy Haven Doe listened to my Biology 2 lectures about DNA, RNA, and Protein Synthesis, she had no intention of making science her lifelong career. Now she was just back in Cambridge following two "boring, boring" years as a graduate student at Yale. No one there would let her search for the chemical essence of still-elusive repressor molecules thought to regulate gene functioning—the next Holy Grail for hotshot molecular biologists to pursue after the Genetic Code. So now she had become a paid research assistant to the lightheartedly arrogant Harvard Junior Fellow Mark Ptashne, then seeking to molecularly identify the repressor that regulates the early life cycle of the bacterial virus λ. Then they were racing Wally Gilbert and Benno Müller-Hill working in my lab space on the floor below to isolate the repressor that controlled the genes involved in sugar lactose metabolism. Wally and Benno, being first in the game, won this race with Mark and Nancy's λ repressor paper being submitted on December 27, 1966, just two months after I submitted Wally and Benno's lactose repressor manuscript. Happily Nancy and Mark won the next lap of the race, being months ahead in showing that repressors act by binding to specific binding sites on DNA molecules.

Cambridge, September 10, 1966

There are new sounds in the air, the songs of instincts. Though the birds no longer proclaim their territory with song, the outdoors is far from quiet.

Chief among these noisemakers are members of the most musical family on earth. Crickets and grasshoppers. All through autumn, these sounds will continue. A light frost will stop a few, but the rest will sing on. Finally the killer freeze will arrive, and they all will be silenced.

Cambridge, September 1966

In the *Christian Science Monitor* today Roger Peterson, the ornithologist, writes on the mounting destructive pressure mankind has brought to bear on wildlife. He discusses all the factors that have caused 80 birds to become extinct in the world, in less than three centuries.

Bulldozers changing the countryside, the doing away with hedges, thickets and ponds, excessive hunting, airplanes, tall buildings and television towers, the dumping of waste oil, and the undisciplined use of residual poisons.

Man has acted as though he owns the earth. He does not, save his home and his tools. He is but the inheritor. Fundamentally, however, the root of the problem is too many people. Only of late is the world tackling the problem of birth control. Perhaps it is too late.

There would be no conservation problem if there were not a population problem. There is no point in ignoring this.

"Man has acted as though he owns the earth. He does not."

Cambridge, September 29, 1966

For Senator Dirkson to filibuster against the Civil Rights Bill one day and then to offer a 'prayer in the schools' amendment to the Constitution the next day is, I believe, an unparalleled display of contradictory attitudes. Of course he is thinking of the votes back home in Illinois.

Such a display may very well injure the faith of many people in our American way of life. What are the Negroes to think of this action which will condemn them to their ghettoes? What are idealistic youth to think of it, and all warm-hearted, generous-minded people?

He turns to religion, which tells us we are our brother's keeper. He is afraid, he says, that often housing may violate the constitution, but he is not afraid to open up the constitution and insert into it religion which should be a private family matter. No wonder Martin Luther King says, "We confront desolate days ahead."

"What are the Negroes to think of this action which will condemn them to their ghettoes? What are idealistic youth to think of it, and all warm-hearted, generous-minded people?"

Cambridge, October 10, 1966

Thanksgiving (as well as New Year's) should be celebrated in October. Late November is a melancholy season, bleak and [?]. Most of the birds are gone and the garden is brown. But early October—October is the climax of the year, a fit season for giving thanks. The swamp maples and the sumac are still afire, the sugar maples blaze their orange glory. The migratory white th. sparrows, Hermit Thrushes and juncos are everywhere and the goldenrod and purple asters hang out their banners and if you know where to find it October is the month for the fringed gentian. The major harvesting is done, and any American with outdoor sense really gives thanks—the maples see to it that now is a special season.

The West has its cottonwoods; Europe has its golden beeches, and the tropics have exotic blooming trees at any time of the year. But we have a triumphant fire—every shade of orange, red and yellow that is all our own.

It is this remembered glory of countryside color that so often makes an American nostalgic abroad. This I have felt the last two years in Gibraltar, Venice, and Southern France.

But what I miss here in Cambridge is the smell of burning leaves in the air. It is now against the law! It is a good smell—the smell that spells the end of the living year.

"But what I miss here in Cambridge is the smell of burning leaves in the air."

Cambridge, October 19, 1966

A cold, disagreeable day—the rain never stopped. At tea-time I thought, as I looked through the side window, of the Hermit Thrush I had seen in previous years back of the garage. At first noticing nothing except the rain and the falling leaves, I was about to turn away when suddenly I observed a moving bird, and there was my thrush.

Naturally, it is very unlikely that it was the same bird, but the day of the month and the place were identical.

The Hermit is the real aristocrat of an aristocratic family. This silent reserve goes well with his unconscious grace of movement. He is never awkward. Some of his poses when he is on the ground are strikingly picturesque. It has an attractive lightness of motion.

The Hermit is easily identified by its habit of lifting its tail especially after a lighting and by its ruffles tail (?). It is the only one of our thrushes which has the tail brighter than the back.

This thrush is the latest migrant in the Fall and the first in the spring, and with us it is silent. But its exquisite song, on its nesting ground in the north woods, is considered by many to be the finest bird of song.

In the winter of 1937–38 my son and I observed a Hermit in Eggers Woods, a forest reserve to the South of Chicago. A bird with an injured wing—unable to fly south.

"In the winter of 1937–38 my son and I observed a Hermit in Eggers Woods."

Cambridge, October 28, 1966

Jim is in New York for a few days—spent the entire day in the house. Over two inches of much needed rain.

A quiet day for me with my stamp collection, but now I am trying to determine the value with the possibility I may sell the whole accumulation. I can determine the price I would have to pay for what I have. What a dealer will give me is another matter.

I can well remember the day many years ago when Uncle Dudley [Dudley Crafts Watson] gave the original collection to my brothers and me.

Much delight I have experienced over the years in adding stamps to the original collection and it is with sorrow I even think of giving up this enjoyable hobby.

But time is running out!

Washington, December 17, 1966

A repulsive and shocking picture in the paper today. Body of a Viet Cong guerrilla being dragged in the dust by an armored carrier. The vehicle was plainly marked US Army 12BX84.

This picture does something terrible to the mind. It is degrading. If this is the image that people around the world have of us, it is little wonder that we are despised and hated by many foreign people.

If this had been a picture of one of our own men being dragged behind a North Vietnam truck it would have cause an uproar. It was taken by a Japanese photographer.

And the picture won a first prize in a photo contest!

"This picture does something terrible to the mind."

Washington, December 10, 1966

Dr. Segal will see me Monday morning. He told Betty common symptom with arteriosclerosis. So that is my trouble. Rather advanced stage. I should guess—more pills.

Dad's stomach, which had bothered him much when I was in grammar school, badly flared up during the winter of 1967. By the time I could fly down to Sarasota, his worst pain was past. Not only was his ulcer beginning to heal but he was still in the care of the cardiologist Robert Windom, M.D. Bob, who had looked after Dad beginning with his first stay in Florida, seemed to know everybody important on the West Coast of Florida. A year later Bob heard me lecture on the double helix at the nearby New College, sited on the former majestic grounds of the Charles Ringling estate, the Ringling Circus family. Then New College had University of Chicago–like "Great Books" courses which had brought Dad into contact with new friends that shared his love of books.

Florida, January 1, 1967

You listen to programs here in Florida that you do not hear up North. I heard one this afternoon. It was addressed to the older people and it was clearly addressed to the fears and insecurity of the retired people. I could not help but think that of the American people the most timid and subject to passion are those—some old, some idle through retirement, who live on fixed incomes from investments. Florida and California attract these people in great numbers and it is a shame to excite them as this man did today. Needless to say he was accusing everyone from the President down of Communism.

The retired people down here naturally gather in tightly knit groups and share their fear with one another. Any fluctuation in the cost of living or change in the tax law or international situation which causes variations in stock prices in the real estate market affects their immediate income with the result that these unhappy people live in a state of persistent apprehension.

This constant state of alarm makes them fair game for the man or group with dictatorial desires. They have only to bring charges, no matter how ridiculous or improbable they are, of plots to disturb the delicate balance of the fixed income in order to arouse fear, which is the mother of ferocity.

The stalking horror is Communism with its threat of confiscation of private wealth, and Socialism which implies that they might be forced to share their wealth with less fortunate citizens. Thus they are frightened.

"This constant state of alarm makes them fair game for the man or group with dictatorial desires."

In early April, Dad was back with Betty and Bob in Washington when I went off during the spring break to Ankara, Turkey, to lecture at the new Middle East Technical University (METU), being sent by the Ford Foundation, which had helped bring METU into existence. Then running the Ford Foundation in Turkey was my former teacher Eugene P. Northrup, who later served as Dean of the College at the U of C. When I went on from Turkey to Beirut to lecture at the American University of Beirut, my Ford Foundation hosts struck me as of less academic inclination.

The year before, when I had spent six weeks in East Africa lecturing at its universities, its intelligent Nairobi office head, Frank Sutton, who earlier had been a Harvard Junior Fellow, told me that the Ford Foundation in helping academic centers through the developing world never let the CIA into their activities. In leaving Beirut, on my final day I was driven over the coastal mountain range to the great Roman Temple at Baalbek. I learned much about the Tapline (Trans-Arabian Pipeline) that brought oil from eastern Iraq to the Mediterranean. I suspected my hosts had two bosses.

In May, Dad again used Italian Line boats to take him to and from the major Mediterranean ports. He by then much liked the Italian fare that had so put him off on his first long tour of Italy with his niece Alice. After he came back to the States, he spent much of the summer at the Harbor View Hotel in Edgartown on Martha's Vineyard. Its large dining room almost invariably attracted one or more guests that Dad could enjoy speaking to. By then my book featuring Francis Crick had not only a new title—*The Double Helix*—but a new publisher —Atheneum of New York. Tom Wilson, who long had been head of Harvard University Press, was moving to Atheneum, knowing that he would soon reach the age of obligatory Harvard retirement. Just weeks earlier Harvard's President, Nathan Pusey, in response to a Francis Crick letter threatening legal action, told Tom to stop its publication. He wrote Tom that Harvard should not be in the middle of a fight between prominent scientists. Before I signed the new contract with Atheneum, I had the manuscript read by the prominent New York City lawyer Ephraim London, then one of the top experts on libel in the United States. To my relief, he said that the book contained no libelous statements. In England the publisher Weidenfeld and Nicolson went ahead after he reassured them that what I wrote about Francis represented healthy reality.

At Sea, May 1967

As is my custom, I walked on deck early this morning, before breakfast. It is very seldom that I meet any of the passengers at that time of day.

One of the nicest moments of sea travel is the moment of reaching the deck early in the morning in fair weather. You come out of sleep, out of the stuffiness of your cabin, into all of the freshness in the world. During the night all has been scrubbed for you, the planks and brass glisten. Nothing can ever seem so clean and fresh again as this empty deck, the tumbling sea, the morning light.

Ah the delight of that turn or two on deck before breakfast. And what a wonderful appetite you have!

October 27 [1967?]

Very much pleased about the probable sales of the Text Book [*Molecular Biology of the Gene*]. I gave my copy to a retired chemistry teacher of New York yesterday and he spoke glowingly of it when he returned it this noon. Was going to purchase a copy as soon as he gets back.

And the new book [*The Double Helix*] has tremendous possibilities if you work hard on it and accept whatever advice HF have to offer. Let them handle sent struct, etc. Your job is to see that the excitement mounts up chapter by chapter and to handle the science as only you know it. Have you ever thot of reconstructing some actual conversations you had with Francis in the manner that mistress of Picasso did in her book? I liked your statement at dinner that evening in Cambridge. What I want to do is to show in detail to a young scientist how an important piece of work was actually done.

Edgartown, Martha's Vineyard, July 1967

There are some places and situations that offer you nothing but despair—you cannot be yourself in them. Everything and everyone there cancels you out. You are alone among the enemy—you seem to have wandered out of your world. To cope with such situations or people, you would have to be born again.

A good part of today was like this—the weather was beastly, fog and rain. The conversations I had were not successful, no meeting of the minds, and I began to talk with Mr. L, a corporation attorney from New York City. What a contrast! He expressed my views on almost everything, well read, with a splendid sense of humor. What a wonderful thing it is to come in such a situation upon a person who thinks as you do. Into their ears you can poor your talk and be understood. What a delight!

"Your job is to see that the excitement mounts up chapter by chapter and to handle the science as only you know it."

Upon returning to Cambridge from Martha's Vineyard, Dad moved into a room in the Hotel Continental whose food he always anticipated. Though he had no current cardiac symptoms, I worried that living by himself at 10-1/2 Appian Way had become too much of a burden on his psyche. Then increasingly joining us for supper was the vivacious, very pretty Radcliffe College junior Elizabeth Lewis, who the academic year before had worked part-time helping me with *Double Helix* matters. Over the summer, she had a resort job in Montana, making me miss much her reassuring cheerful presence. To my relief, she was again assisting my secretary Susie Aldrich, often staying on 'til near supper time when we would walk with her bike to join Dad in the old-fashioned basement restaurant, only a several-minute bike ride along Garden Street away from her Radcliffe dorm.

Elizabeth Lewis in 1967.

Cambridge, August 28, 1967

I could hardly believe! There on the front page of yesterday's Sunday edition of *The New York Times* was a picture of some birds long thought to be extinct, the ivory billed woodpeckers.

The *Times* is to be congratulated in daring to give such news as this front page notice when many readers might be of the opinion that world news should have been given the space.

For me it started the day off just right. It gave hope that in this world of turmoil, lost beauty has been found again.

Cambridge, September 1967

Nowadays we have the Television of cruelty. Inasmuch as there is a set in my room at the Continental, I watched it a little. I'll not do it very much from now on for several reasons, one of which are their shots of the war.

NBC or CBS has not got around to throwing communists to the lions in front of the cameras, but it has got Vietnam, where so-called communists are slaughtered in more modern ways.

Vietnam is the most televised war in history, a sort of 24 hour theatre of cruelty, endlessly lulling the Americans from their real problems.

Meanwhile, the yellow and black and brown hordes are getting more hungry and restive, waiting at the games. The parallels between ancient Rome and us are very interesting.

Cambridge, September 1967

September is here and I make the attempt to adjust to another hotel—my memories go back to other Septembers.

September, of course, is always best in the region where you spent your youth and the greater part of your life in my case, the Midwest, Chicago, and the Indiana Dunes.

I think of those Septembers when I was afield either with the children or their mother; September fire-red or yellow gold as the leaves began slowly to turn the mouth with the taste of summer at one end and the taste of cold at the other.

Yet September in the Midwest with all its wetness carried melancholy with it and I remember other Septembers in other places where sometimes there was elation and hope.

September at Oberlin when I was away from home for the first time with all the excitement of the first year at college.

September on the western front in World War I when we were being transferred from the British front to the newly formed American army—all of us filled with the hope we were doing our bit to make the world safe for democracy!

September in Paris in 1959, a lovely month with gaiety and life returning to the boulevards and the August refugees from the countryside again settling down in the sidewalk cafes and bistros for the blue hour, that mysterious, quiet prologue to the wonders of the Parisian night, chestnuts on the charcoal, braziers filling the air their special autumn tang, and the oyster openers (?) outside of the cafes and restaurants.

And September in the far-east and the thrilling trip to Boston from Phnom Penh, through Bangkok, Hong Kong, Tokyo, and San Francisco.

And finally September here in Cambridge which is simply people charged for action, a month of hope, the season to try again.

"My memories go back to other Septembers."

"September in Paris in 1959, a lovely month with gaiety and life returning."

Cambridge, September 26, 1967

After reading *The Territorial Imperative* by Robert Ardrey and *On Aggression* by Konrad Lorenz, I feel that to considerable extent we are still uncertain about the basis for aggression in human behavior.

Both authors declare that fighting is an instinct which demands expression in virtually all high animals and also in ourselves because we have inherited this compulsion. It is in our genes. They have been answered by various writers who state that Ardrey and Lorenz have misinterpreted their data.

Be that as it may I think a good case could be made for the positions that the danger of war comes from man's good instincts, and not from his bad.

We see that right through history men have been willing to die for the things they value: their women and children, their land, their country, their religion, their ideas.

What we have to worry about is not simply aggression. The will to power, destructiveness, violence—but man's age long habit of dying for values he cares more about than himself.[3]

January 8, 1968
The Anchorage
Nokomis, Florida

Half a dozen of the guests here have read the *Atlantic* installment. They don't understand the science but have found the story interesting and they all speak of how well it is written. Cannot find a line in the book that sounds like it was written by any one else—all to your credit.

Do try however in your interviews to make the story as simple as you can. There are many older people like myself who have difficulty in understanding the chemistry but these are the people who have the money to buy the book!

When the reviews of the book start coming in, I'll see a few of them. I would like to do a scrapbook as I did with the Nobel but I'll have to depend on you to save the reviews. It will give me something to do when I come north and you will find it interesting in the future.

[3] JDW, Sr., had actually met Robert Ardrey on one of his European trips. In an undated letter written to JDW, Jr., upon reaching Trieste, he related, "One man, Robert Ardrey, living in Rome, wanted to talk to you—a brilliant chap. Been working in Hollywood for some years—almost won an Oscar this last month. Evidently has earned considerable money and from now on want [*sic*] to write on serious matters and is interested in microbiology. He recently published *The Territorial Imperative* [in 1966], which received fine reviews—a graduate in anthropology at Chicago." Ardrey in his introduction to the *The Territorial Imperative* described the book as bringing "into focus a single aspect of human behavior which I believe to be characteristic of our species as a whole, to be shaped but not determined by environment and experience, and to be a consequence not of human choice but of evolutionary inheritance."

THE *Atlantic*

The Race to Discover the Secret of DNA

by *James D. Watson*

Beginning: **The Double Helix,** a book about the biggest event in biology since Darwin

Cover of the first installment of The Double Helix *in* The Atlantic Monthly.

January 15, 1968
The Anchorage
Nokomis, Florida

   Bill Kennedy has just flown in from Zurich—carried a copy of *The Atlantic Monthly*. He was interested in your treatment of Rosy Franklin—he thot that she came to life very well.

Comments from Linton Keith—

   I am glad you told me about the serialization of the *Double Helix*—I immediately went out for the January issue and read the first installment with growing interest and delight—although it may be somewhat lacking in the grace of good style that qualifies a writing as 'literature' for the kind of writing that it is meant to be, it has a unique personal style that adds fascination to its subject.

Cover of the first edition of The Double Helix, *which was published in 1968 after being serialized in two early-in-the-year issues of* The Atlantic Monthly *magazine. The book had been offered to Harvard University Press in the fall of 1966. The Harvard University Board of Syndics, composed of 12 advisory professors, recommended publication, but they were overruled by the Harvard Corporation. Francis Crick had protested publication, writing that "you have grossly invaded my privacy," and Maurice Wilkins felt that* The Double Helix *painted "a distorted and unfavourable image of scientists." Pusey's position was that publishing the book "would be to take sides in a scientific dispute." The Harvard Crimson jumped into the fray, saying that the reputation of Harvard University Press had "probably" been jeopardized by this decision, and accused Pusey of being "less interested in diversity of viewpoint than bland tranquility." But controversy or no, the book was an immediate success, gaining spots on best-seller lists worldwide.*

Already then Harvard University had agreed to let me assume as of February 1, 1968, the Directorship of the Cold Spring Harbor Laboratory. Long the emotional home of American genetics, I had much admired its scientific psyche ever since I spent the summer of 1948 there doing research and much enjoying its bucolic waterside location. Its financial stability had been long stabilized by funds coming from the Carnegie Institute of Washington. Their 1963 decision to withdraw their support from Cold Spring Harbor soon lessened its capacity to attract scientists of the first rank. Though most of my time and my main residence would be spent at Harvard, I thought that by having a first-class Assistant Director on my side I could successfully change the Lab's direction from bacterial cells to those of higher vertebrates, in particular, those of human cancer cells.

Dad by then had stopped worrying that *The Double Helix* would harm my reputation and keenly looked forward to its serialization in the January and February editions of *The Atlantic Monthly*.[4] When Dad joined Betty and Bob in Washington for the holidays, he had already seen an advance copy of the January issue. My Christmas 1967 plans were up in the air until the Czech-born Nobel Prize–winning biochemist Carl Cori and American second wife Anne asked me to join them at the old-fashioned 'Tween Waters Inn Island Resort on Captiva Island, just south of where Dad spent his winters at Nokomis. By mid-February, the first book reviews were appearing with the most influential that in *The New York Times Book Review* by Columbia University's prominent sociologist Robert Merton. He wrote that *The Double Helix* was a candid self-portrayal of the scientist as a young man in a hurry. Within a month it was on *The New York Times* best-seller list, staying there for 16 weeks, briefly reaching the #5 spot.

As soon as a new review would appear, I had Liz send a copy to Dad, whose Christmas-time cough had not abated but steadily worsened. By mid-February he was in the hands of his savior of the winter before, Bob Windom. But this time Bob could not help him. The X rays taken to

[4] The book is dedicated to Naomi Mitchison. In a March 25, 1968, to JDW, Jr., she wrote: "It's a great thrill seeing Honest Jim (though I think the present is the more definitely romantic title—it sounds like a Celtic fairy tale) actually in print. I've seen quite a lot of reviews including that smashing one in *The Scientist* and I feel almost as if it was *my* child." She added, "I do hope you and Francis are not fighting. It seems a bit silly. There are so many real things to fight about and Francis comes out pretty well, considering."

JDW, Sr., in Florida, 1967.

possibly reveal pneumonia instead displayed the telltale marks of widely disseminated lung cancer. Betty was the first to learn of Dad's incurable cancer from Bob Windom. That day I was returning from New York where I helped share the first hour of *The Today Show* with Harry Belafonte, only getting back to my office late in the afternoon. By then Susie had left, with Liz answering Betty's call telling me that Dad's future life would be measured in months.[5]

Liz and I went straight to the Continental Hotel dining room to let us ponder over what I should do next. Afterward she came back with me to Appian Way, staying for the night and the next day moving effectively permanently out of her Radcliffe dorm room. Since that night we have been together for 46 consecutive years. After taking her the next evening to a book party at the Cori house, I flew directly to Sarasota to bring Dad back to Betty and Bob's. They, in turn, put him under the care of local Washington cancer doctors, who started treatment with the selective cancer cell killer nitrogen mustard. Then, as still today, such chemotherapy at best delays but never cures those so treated.

From virtually our first night together, Liz and I saw ourselves soon marrying, possibly during my forthcoming trip to San Diego, where I would be attending a late March meeting bringing science writers together with science focusing on studying and treating cancer. I was an obvious person to invite since as the new Director of the Cold Spring Harbor Laboratory I intended to make its main focus how tumor viruses convert normal cells into cancer cells. Immediately following the conference, the weeklong Harvard spring break was to begin, letting Liz fly from Boston to San Diego on March 28, 1968. That evening at 9:00 p.m., we were married at the Congregational Church sited in the center of La Jolla, San Diego's posh northern enclave. There, directly over-

Doctor and Mrs. Robert V. Lewis
have the honour of announcing
the marriage of their daughter
Elizabeth
to
James Dewey Watson
on Thursday, the twenty-eighth of March
One thousand nine hundred and sixty-eight
Union Congregational Church
La Jolla, California

*Announcement of Jim and Liz's
wedding.*

Liz and Jim at their March 28, 1968, wedding. Nancy Hopkins, long a correspondent with JDW, Sr., wrote to him that, "I'm not sure that my emotional makeup could take another week like this. First LBJ—No sooner had we opened the champagne to celebrate than Jim dropped his bombshell. As far as the Harvard Biological Labs were concerned, JDW more than took the wind out of LBJ's sails. I never knew that a marriage could cause such pandemonium—people were screaming, beaming, literally jumping around with excitement. To sum up the response in very mild terms I can only say that everyone considered it a wonderful choice, wonderful event—executed with true flair. You must be very happy indeed."

[6] CBS news anchor Walter Cronkite in a special report on Vietnam on February 27, 1968, had said that, "It seems more certain than ever that the bloody experience of Vietnam is to end in a stalemate.... It is increasingly clear to this reporter that the only way out will be to negotiate, not as victors, but as an honorable people who lived up to their pledge to defend democracy, and did the best they could." With the rising tide of protest and now this, Johnson is said to have remarked, "That's it. If I've lost Cronkite, I've lost middle America." On March 31, 1968, Johnson announced that, "We are prepared to move immediately toward peace through negotiations," that he was substantially reducing the "present level of hostilities ... unilaterally," and finally that he would not seek reelection.

looking the Pacific Ocean, Jonas Salk had wisely sited his new institute across from the proposed new campus of the University of California. The next day we drove to the Borrego Springs on our way to the Mexican fishing village of San Felipe at the top of the Gulf of California. On our drive from there to Tucson, our car radio let us learn that Lyndon Johnson had just announced that he would not run again for President.[6] To his profound dismay, he saw himself presiding over an ever-growing, never-winning American military presence in Vietnam. It so beat

Scientist Begins His and Lab's New Life

By David Zimman

Cold Spring Harbor—Dr. James D. Watson, the Nobel Prize-winning biochemist who is 40 years old today, returned yesterday from a brief Mexican honeymoon with his 19-year-old bride and could not resist showing off his laboratory here.

Watson, a Harvard professor and author of the controversial best-selling scientific autobiography "The Double Helix," became director on Feb. 1 of the renowned but deteriorating Cold Spring Harbor Laboratory of Quantitative Biology. The small lab, which has been near bankruptcy for the past 10 years, has concentrated on genetics. But Watson plans to launch a five-year, $3,000,000 fund-raising campaign and turn the laboratory's emphasis toward basic cancer research.

Radcliffe Junior

"I don't pretend to understand all his work," said the former Elizabeth Lewis of Providence, R.I., a pretty, green-eyed Radcliffe junior. She spoke as the strolled through the lab's wooded 35 acres and by its 30 buildings, some dating back to the early 19th Century. "But I think we get along beautifully despite the difference in our ages. Jim is a youthful person. He's forward-thinking and he's boyish-looking. But he's too busy with his work he needs someone to take care of the little things that are important at home," Mrs. Watson, who had a part-time job in her husband's Harvard office, married him March 28 in La Jolla, Calif. He was addressing a cancer seminar there.

Watson, a slender man who constantly runs his hands through his thinning brown hair while he's talking, won the Nobel Prize for medicine in 1962 with two British scientists. The trio discovered the structure of DNA, the molecule that governs heredity.

In an interview yesterday, Watson said he planned to devote his summers to the lab. In the winter, he said, he would spend about two-thirds of his time at Harvard and one-third at Cold Spring Harbor. "The lab will be increasingly emphasizing research on animal cells and viruses, Watson said, "and I hope it will become a major center for basic cancer research.

While viruses have been found to cause cancer in animals, no human cancer has yet been traced to a virus. However, some investigators, including Watson, believe viruses cause some forms of cancer. "The problem is you just can't experiment with humans," he said. "But it is so likely I don't have any doubt . . . The challenge is to find what transforms a cell from normal to cancerous."

The laboratory, founded in 1894 on a quiet North Shore cove, is famous for its summer symposiums and teaching programs that attract world-famous scientists. In addition, its year-round staff has made key contributions in the development of hybrid corn, in fruitfly genetics, and, most recently, in the genetics of bacterial viruses and bacterial resistance to antibiotics.

But the laboratory has had a hand-to-mouth existence on research grants. It has virtually no endowment to use to maintain its $3,000,000 plant or to pay attractive staff salaries. Consequently, Watson said, it has not been able to expand or maintain its buildings and grounds adequately.

Watson's rejuvenation plans call for a $2,000,000 cancer research building, a $400,000 library, tripling the present staff of 12 researchers with doctorates, and doubling its summer student enrollment by enlarging the dormitories.

GEORGE LANE NICHOLS MEMORIAL

Saturday, April 6, 1968

Newsday

Watson Shows His Bride the Laboratory Yesterday

Newsday *article—"Scientist begins his and Lab's new life."*

---

down his Presidency that his much-hoped-for Great Society's "War on Poverty" had no chance to move forward.

Upon our arrival in Washington, we found Dad still just moving about and enjoying seeing me married to someone who fulfilled all my past dreams. After three days, Liz and I drove up to Cold Spring Harbor, some 30 miles east of New York City on Long Island's North Shore. Then for a day I showed Liz the early 1800s Osterhout Cottage where we would live while we started regularly coming down from Harvard. That night an impromptu dinner party was held to celebrate our marriage. Its mood could not be totally joyous. Late in the morning of April 4, 1967, Martin Luther King, Jr., was assassinated in Memphis.

Our next visit to Washington found Dad in nearby Sibley Memorial Hospital to start medicines to bring his pain and general discomfort more under control. In early June, when Liz's Harvard exams were over, we returned to Cold Spring Harbor, where soon a large house on

Shore Road was rented to let Betty bring Dad up to Cold Spring Harbor. Before their arrival, a second celebratory dinner party was arranged—one marking my assuming the Directorship and hopefully soon bringing the Lab's financial distress to an end. Again, its final mood was not what was anticipated. Bobby Kennedy earlier that day in the early-morning hours of June 5th had been assassinated in Los Angeles while mounting his campaign for Presidency. I had much wanted him as the Democratic candidate. He had the guts to stop a war that our nation would never win. To my dismay, the only viable Democrat remaining as a candidate was Senator Hubert Humphrey, whom I saw as more a man of speeches than of hard decision making. Getting our nation soon out of Vietnam had become an unrealistic immediate objective.

That summer I brought with me to Cold Spring Harbor eight younger scientists from Harvard studying phage λ. Nancy Doe Hopkins came

Liz and Jim at Cold Spring Harbor in 1968.

down with Brook for several weeks, allowing them on several occasions to see Dad, who steadily was losing his strength. Now his doctor was the warmly intelligent College of Physicians and Surgeons–trained internist Reese Fell Alsop, who like many of the Lab's immediate neighbors had been an undergraduate at Harvard. From the start we required daytime nursing help that soon became 24-hour care. As July came toward its end, Reese urged us to move Dad to the nearby Hilaire Nursing Home, where Dad's longtime favorite socialist Norman Thomas had just spent the last years of his life. Dad's stay at Hilaire mercifully lasted only a day. On July 29, 1968, he died at the age of 70.

Betty and I flew separately to Indianapolis from which we drove up north 120 miles to Chesterton for the burial of Dad's ashes in a cemetery plot next to where our mother Jean had been buried 11 years before. Mother's cousins came from Michigan City to join a very small group of Watson family friends all saddened by the passing of a kind man whose long devotion to our family carried us through the Great Depression and World War II, allowing us to participate in the ever-increasing prosperity of America's first postwar decades. Afterward Betty and I drove our rental car into Chicago so that we could look again at the Luella Avenue bungalow where our youths were passed. Then we drove toward the Loop, easily passing through Grant Park, where the day before all-too-young Americans had been brutally clubbed by the police for demonstrating against the United States' involvement in Vietnam. Its broad lawns were now almost ghostly as the delegates to the Democratic Convention were about to meet to nominate Senator Humphrey as their party's Presidential candidate.

If he had not begun to smoke at Oberlin and all through his World War I time in France, Dad might well have lived, like his brother Bill, into his early nineties. Then he would have had the joy of seeing that both Betty and I had chosen mates who not only shared his passion for books but who also took pleasure in creating new ones. I now look back to my good fortune in having Dad pass on to me his lifelong quest for truth, reason, and decency.

Afterword

In looking back to what factors most made possible my unquestionable position as one of today's most successful biologists, I rank the most important my close relationship with my Dad. It started in 1938 when at age 10 my legs had become sufficiently long to let me routinely accompany him on bird-watching trips to nearby Jackson Park. While my subsequent university education at the University of Chicago made me start serious thinking, it could not have given me by itself the intellectual and emotional depth of a serious 25- (if not 30-) year old when I was only 20. Talking to Dad about serious adult matters was even more important. So while I grew up without a trace of any of the childhood precocity that very early on dominated the life of my distant cousin, Orson Welles, by the end of my first year as a graduate student at Indiana University I had all the essential trappings of the mature intellectual. I knew what was important and as much as possible did not spend even brief moments with those of lesser intellectual bent.

Growing up during the economic depression of the 1930s and watching the rise of Adolf Hitler likely propelled me into the ranks of serious adulthood much faster than if I had grown up during the earlier Harding, Coolidge, and Hoover presidencies. But that a Franklin Roosevelt presidency was a necessity and not a luxury for the United States' future would not have so dominated my adolescent days had not Dad passed on to me the societal goals of his once well-to-do father's Republican Party's last great president Theodore Roosevelt: The rich had to begin being taxed to redistribute monies to the poor, the monopolistic Standard Oil Trust had to be broken up into truly competing corporations, and our country's vast wilderness regions had to be saved for all future Americans. Most important to Dad were the nearby Indiana Dunes National Lakeshore in which the rare Piping Plover still nested.

Dad in no way tried to stop me when I would argue with any adult who revealed himself as a Franklin Roosevelt hater or Hitler-favoring isolationist. The world was in too bad a shape to let such nonsense go unquestioned. Listening to the evening's radio wartime news was always of high priority. Evening meals never occurred at the expense of not knowing as soon as possible who was winning the latest battle. By the age of 11, I had stopped going to the Catholic masses of my grandmother knowing by then that Dad did not believe in any form of God. Soon he also let me know there was no value gained by arguing with either school friends or adults about religion even though there was no evidence for the existence of gods of any form. As the overwhelming majority of Americans still so believed in God, there was no point of ever directing my life into the public arena of politics where godless societies were spoken of with horror and disgust.

Just as I now have no memories of ever even mildly disagreeing with Dad about anything important, I remember equally well much liking my mother, who only on Easter or Christmas went to Mass and was never dominated by equally irrational pursuits like astrology or quack medicines. But in contrast to Dad, who clearly preferred reading about ideas than talking to ordinary people, Mother was inherently a people person with the capacity to politely put up with all Americans be they Democrat or Republican or educated or uneducated. Clearly making my mother much liked was her stable optimism about the future. In this regard, I was also more emotionally like my mother than my dad for whom bad news frequently led to lengthy depressive thoughts. My unanticipated, shockingly early big success with DNA, of course, much reassured Dad and Mother. After Mother died in 1957 of a childhood rheumatic fever–damaged heart, I was by then making sufficient money to help Dad live a modestly comfortable life. My sister's marriage also had reason to give him much comfort. Her husband, Bob Myers, was not, as we earlier feared, a Republican but a well-educated, Indiana-born Democrat immersed in the reality of high-powered foreign policy.

How much I also owe my success to the fact that my ancestors likely made many more good marriages than bad and were inherently much more optimistic about their futures than the facts of the time dictated is not now answerable. On the other hand, the writings of my father as evidenced through this book's pages leave little doubt of his high intelligence and ability to write clearly in the English language. Most certainly I didn't emerge from nowhere!

Jim Watson
21 February 2014

APPENDIX 1

James Dewey Watson Genealogy

This description provides sources for lineages of the paternal and maternal family lines of James D. Watson shown in the pedigree chart. On the paternal side, the parentage of William Weldon Watson (1794–1874) is unknown, as is that of his first wife Maria Cape Humerickhouse (1793–1834). There are, however, two ancestors on the paternal side from Puritan families that have well-documented ancestries that date back to 11th-century England. They are Augusta Crafts Tolman (m. William Weldon Watson III, parents of Thomas Tolman Watson) and Sarah Loomis Dewey (m. Delavan Ford, parents of Nellie Dewey Ford). From these two women we can trace James D. Watson back to several early New England settlers, including the Bingham, Chandler, Clark, Crafts, Loomis, Cheney, Hunt, Paddock, Phelps, Ingersoll, and Strong families. More distant notable ancestors include Thomas Dudley, John Winthrop, Sir John Popham, Cardinal Henry de Beaufort, John of Gaunt, and King Edward III. The lineages of the maternal Mitchell and Gleason families remain murky because they were relatively recent immigrants to America and their fairly common names make it difficult to distinguish the relevant ancestors.

The lineages of two prominent people in this book and their relationship to James D. Watson are as follows. *Dudley Crafts Watson* (1885–1972) was the son of Augusta Crafts Tolman Watson (1852–1927) and William Weldon Watson III (1847–1871). He was the brother of Thomas Tolman Watson (1876–1930), James D. Watson's grandfather. *Orson Welles* (1915–1985) is descended from Abigail Watson (daughter of Benjamin A. Watson, the father of William Weldon Watson III) and

The original version of this genealogy was prepared by John Zarrillo, James D. Watson Archivist, Cold Spring Harbor Laboratory Archives.

A complete genealogy can be found at http://watson.ancestor-tree.net.

B: Born
P: Place
M: Married
D: Died

2 James Dewey Watson, Sr.
B: 16 Dec 1897
P: Eveleth, MN
M: 23 May 1925
D: 1968
P: Huntington, NY

4 Thomas Tolman Watson
B: 16 Aug 1876
P: Washington Heights, IL
M: 1895
P:
D: 20 Mar 1930
P:

5 Nelly Dewey Ford
B: 14 May 1875
P: Wisconsin
M:
P:
D: 3 Dec 1936
P: Chicago, IL

8 William Weldon Watson III
B: May 1847
P: Illinois
M: 25 Oct 1871
P: Kane County, IL
D: 1913
P:

9 Augusta Crafts Tolman
B: 10 Jun 1852
P: Portland, ME
M:
P:
D: 1927
P:

10 Delavan Ford
B: 1825 or 1830/1832
P: New York
M: Bef 1870
P:
D: 29 Nov 1905
P: Milwaukee County, WI

11 Sarah Loomis Dewey
B: 6 Feb 1839
P: Cooperstown, NY
M:
P:
D:
P:

16 Benjamin A. Watson
B: 9 Dec 1818
P:
M: 11 Feb 1845
P:
D: 21 Jun 1901
P:

17 Emily R. Planck
B: 1827 or 1828
M:
P:
D: 28 Jul 1871
P:

18 Thomas Frothington Tolman
B: 2 Dec 1819
M: 31 Jul 1850
P:
D:
P:

19 Augusta Maria Paddock
B: 30 May 1826
M:
P:
D: 11 May 1885
P:

20

21

22 James Johnson Dewey
B: 14 Aug 1814
M: 13 Apr 1838
P:
D: 29 Mar 1898
P:

23 Eliza Ann Bates
B: 1816
M:
P:
D: 22 Aug 1858
P:

32 William Weldon Watson
B: 1 Apr 1794
P: Sussex County, NJ
M: 15 Mar 1818
P: Lexington, KY
D: Nov 1874
P:

33 Maria Cape Humerickhouse
B: Jan 1793
P: Lexington, KY?
M:
P:
D: 17 Jul 1834
P: Nashville, TN

34 Sarah Wiley Mottashed
B:
P:
M: 19 May 1842
D:
P:

35 Jacob C. Planck
B: 1804
M:
P:
D: 1867
P:

36 Mary B. Rogers
B: 1806
M:
P:
D: 1867
P:

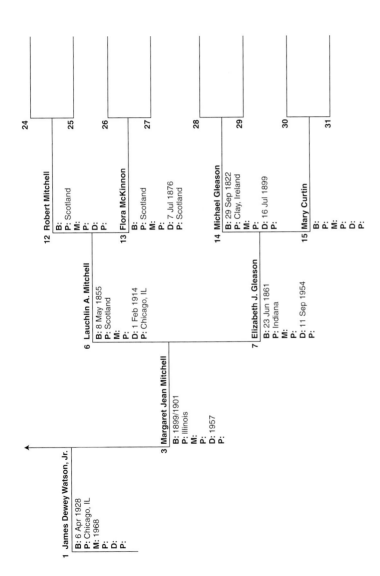

1 James Dewey Watson, Jr.
B: 6 Apr 1928
P: Chicago, IL
M: 1968
P:
D:
P:

3 Margaret Jean Mitchell
B: 1899/1901
P: Illinois
M:
P:
D: 1957
P:

6 Lauchlin A. Mitchell
B: 8 May 1855
P: Scotland
M:
P:
D: 1 Feb 1914
P: Chicago, IL

7 Elizabeth J. Gleason
B: 23 Jun 1861
P: Indiana
M:
P:
D: 11 Sep 1954
P:

12 Robert Mitchell
B:
P: Scotland
M:
P:
D:

13 Flora McKinnon
B:
P: Scotland
M:
P:
D: 7 Jul 1876
P: Scotland

14 Michael Gleason
B: 29 Sep 1822
P: Clay, Ireland
M:
P:
D: 16 Jul 1899
P:

15 Mary Curtin
B:
P:
M:
P:
D:
P:

24

25

26

27

28

29

30

31

John G. Ives. Welles' mother Beatrice Ives Welles (1882–1924) was the granddaughter of John G. Ives.

First Generation (Great-Great-Great Grandparents)

William Weldon Watson: Almost all information regarding WWW comes from the *History of the Early Settlers of Sangamon County Illinois* by John Carroll Power (1876). WWW was born 1 April 1794 in Sussex County, NJ. He then began moving west, first to Lexington, Kentucky where he married *Maria Cape Humerickhouse* (Cape being her maiden name) and from there to Nashville, Tennessee. Here they had five or six children (Benjamin, Abigail, Ann Maria, Hester, and Cordelia are listed in the *History of the Early Settlers of Sangamon County*). Notes from a previous researcher indicate that William and Maria's oldest child was a William Weldon Watson, Jr. This could explain why Benjamin's son was designated William Weldon Watson III; however, no records have been found to substantiate this. Mrs. Maria Watson died 17 July 1834 and WWW subsequently moved to St. Louis, Missouri where he was in business for two years. Finally he settled in Springfield, Illinois. On 19 May 1842 he married a widow, Sarah Wiley Mottashed. According to the 1860 Federal Census he was still living in Springfield and working as a confectionary. This Census lists his birthplace as New Jersey.

One final source of information is from a compendium of legal filings entitled *Private Laws of the State of Illinois Passed by the Twenty-Fifth General Assembly, Convened January 7, 1867. Vol. II*. In March 1867 an act incorporating the Perry Springs Hotel and Railroad Company was approved. Among the businessmen who were named as having an interest in the corporation were W.W. Watson and B.A. Watson, of Illinois. For more information about this business, see the entry on Benjamin A. Watson.

Thomas Tolman was born in Attleboro, Massachusetts on 22 April 1786 (1850 Federal Census). He was the son of Lt. Thomas Tolman and Lois

Clark. He married Jane A. Cook on 25 October 1814 in Arlington, Vermont. By 1819 the family was living in Craftsbury, Vermont. According to the 1840 Federal Census, Thomas was a veteran, and at the time living in Greensboro, Vermont. He had six children. He died on 27 April 1856 in Portland, Maine. He is buried in Evergreen Cemetery in Dearing, Maine (http://thomas.tolmanfamily.org).

Jane A. Cook was born in Vermont in 1792. Her parents were Charles Cook and Elizabeth Burbeck (both of Newbury, Massachusetts). The family moved to Greensboro, Vermont in 1801. She married Thomas Tolman on 25 October 1814 in Arlington, Vermont. She had six children. She died 11 September 1860 in Portland, Maine and is buried in Dearing, Maine (http://thomas.tolmanfamily.org).

William E. Paddock was born in Vermont in 1800. His parents were James Paddock and Augusta Crafts. By 1826 he was married to Mary Clark and living in Craftsbury, Vermont. In 1850 he was still in Craftsbury and working as a merchant (1850 Federal Census). He died around 1855.

Mary Clark was born in New Hampshire in 1802 (1850 Federal Census). By 1826 she was married to William E. Paddock and living in Craftsbury, Vermont. She was still residing in Craftsbury in 1850.

Second Generation (Great-Great Grandparents)

Benjamin A. Watson: *History of Early Settlers of Sangamon County Illinois* records Benjamin's birth as 9 December 1818 in Nashville, Tennessee. He was the son of WWW and Maria (Cape) Watson. He married *Emily R. Planck* 11 February 1845 in Springfield, Illinois. They had seven living children. This history lists his father as W.W. (of New Jersey) and Maria (Cape) Watson (of Kentucky). According to her death announcement in a local paper, Emily died 27 July 1871 in Perry Springs, Illinois.

Another source of information on Benjamin is *The Pike County History*, which was written by Jess. M. Thompson between 1935 and 1939 and was published in 1967. According to the book's Foreword, "[he] was a contributing editor of the Pike County Republican, a weekly newspaper published at Pittsfield, Pike County, Illinois." For the most part the sources seem to confirm each other. According to the history, Benjamin was born in 1818 in Nashville, Tennessee. He immigrated to Illinois when he was 18 years old. In 1845 he married Emma R. Planck. The history confirms that the couple had seven children. It also states that in 1865 Benjamin left Springfield, Illinois and the confectionary business to establish a resort at Perry Springs, Illinois (the Perry Springs Hotel and Railroad Company, see the previous entry on WWW). Mrs. Watson died in 1870. Benjamin also served as the postmaster in Perry Springs.

There are two additional sources of information. The *Illinois Marriages, 1790–1860* online database (provided by http://www.ancestry.com) provides the following information: Emily R. Planck married Benjamin A. Watson on 11 February 1845 in Sangamon Count, Illinois. The final source of information comes from the 1900 Federal Census. In 1900 Benjamin was 81 years old and living in Chicago, Illinois. His birth is dated Dec 1818 in Tennessee. His father's birthplace is listed as New Jersey and his mother's as Kentucky. Two of his daughters, Hester and Annie [?], resided with him. Their mother's birthplace is listed as Illinois. A record from the *Illinois Statewide Marriage Index, 1763–1900* confirms this marriage date and location.

Thomas Frothington Tolman was born in Craftsbury, Vermont on 2 December 1819. His parents were Thomas Tolman and Jane A. Cook. He married Augusta Maria Paddock on 31 July 1850 in Craftsbury, Vermony (*The Craft Family: A Genealogical and Biographical History*). The family then moved to Portland, Maine, where they remained in 1860 (1850 and 1860 Federal Censuses). His only child, Augusta Crafts Tolman, was born on 10 June 1852 in Portland according to the Maine Birth Records of the City Clerk's Office, Portland. By 1880 he was living in Lake Geneva, Wisconsin. Here he worked for his son-in-law, William Weldon Watson III, at the Whiting House Hotel (1880 Federal Census).

Augusta Maria Paddock was born on 30 May 1826 in Craftsbury, Vermont (*The Craft Family: A Genealogical and Biographical History*). She married Thomas Frothington Tolman on 31 July 1850 in Craftsbury. The family then moved to Portland, Maine, where they remained in 1860 (1850 and 1860 Federal Censuses). Her only child, Augusta Crafts Tolman, was born in 1852 in Portland. By 1880 Augusta Maria was living in Lake Geneva, Wisconsin (1880 Federal Census). She died on 11 May 1885 (*The Craft Family: A Genealogical and Biographical History*).

James Johnson Dewey was born in DeKalb, New York on 14 August 1814 (according to the *Life of George Dewey...*). His parents were Chester Dewey and Mary Ann Johnson. He married Eliza Ann Bates on 13 April 1838 in Cooperstown, New York. *The History of Walworth County, WI* accounts for much of James' life after he moved to Lake Geneva, Wisconsin. In 1845 opened a hat store in Geneva, Wisconsin and served as postmaster in 1849. Later he served as President of the Village and Associate Supervisor. Eliza died 22 August 1858 in Lake Geneva of consumption. James was still living in Lake Geneva in 1880 (1880 Federal Census). He later married Selina A. Merriam (1827–1870). He died on 29 March 1898 in Lake Geneva, Wisconsin.

Eliza Ann Bates was born in Amenia, New York in 1816. She married James Johnson Dewey on 13 April 1838 in Cooperstown, New York. By 1845 the family had moved to Lake Geneva, Wisconsin. She died on 22 August 1858 in Lake Geneva of consumption. The *Life of George Dewey* is the only record of Eliza available.

Third Generation (Great Grandparents)

William Weldon Watson III was born in May 1847 in Springfield, Illinois to Benjamin A. Watson and Emily R. Planck. He went on to marry Augusta Crafts Tolman (1852–1927).

Sources of information on WWWIII include a marriage record found in the online database (provided by http://www.ancestry.com) *Illinois Statewide Marriage Index, 1763–1900,* indicating that WWWIII married Augusta Crafts Tolman on 25 October 1871 in Kane County, Illinois. Augusta Crafts Tolman is the ancestor of several notable Puritan families, including the Crafts, Paddocks, Chandlers, and Dudleys. Through Anne Dudley the Watson family can be traced back to two Governors of the original Massachusetts Bay Colony, Thomas Dudley and John Winthrop. The 1880 Federal Census confirms that he lived in Lake Geneva and ran a hotel. According to the 1900 Federal Census he had moved to Chicago, Illinois and was in the real estate business. Census records confirm his birth date and place of birth. He and Augusta had four sons—WWWIV (1872–1958), Thomas Tolman (1876–1930), Dudley Crafts (1885–1972), and Garnett (1891–1959).

Augusta Crafts Tolman was born on 10 June 1852 in Portland, Maine. Her parents were Thomas Frothington Tolman and Augusta Maria Paddock. According to the 1860 Federal Census the family still lived in Portland, Maine. She married William Weldon Watson III on 25 October 1871 in Kane County, Illinois. By 1879 she had relocated, along with her husband, children, and parents, to Lake Geneva, Wisconsin. In 1900 she lived in Chicago with her husband and sons. She died in 1927.

Delavan Ford was born in New York (1870 Federal Census). His parents were both born in New York, although there is no record of their names. According to the 1870 Federal Census he had moved to Bloomfield, Wisconsin and married Sarah L. Dewey by 1870. There he lived with the Dewey family. By 1880 the family had relocated to Lake Geneva, Wisconsin and he was employed as a lumber dealer. He lived with his wife, two daughters, and father-in-law James J. Dewey (1880 Federal Census).

Sarah Loomis Dewey was born in Cooperstown, New York on 6 February 1839 (1870 Federal Census). According to various censuses she had moved to Lake Geneva area by 1845 and remained there for the rest of the 19th century. She married Delavan Ford in Wisconsin sometime before 1870. Her death date and place remain unknown.

Fourth Generation (Grandparents)

Thomas Tolman Watson was born on 16 August 1876 according to a registration card, and his birth place was listed as Washington Heights, Illinois. His parents were WWWIII and Augusta Crafts Tolman. The family moved to Lake Geneva by 1879.

He married Nellie Dewey Ford around 1895. According to the 1920 Federal Census Thomas Tolman Watson was living in Chicago, Illinois with his wife and four sons (Stanley Ford, William Weldon, James Dewey, and Thomas Tolman II) and was employed as a stockbroker.

Nellie Dewey Ford was born in 1875 in Lake Geneva, Wisconsin (according to the 1930 Federal Census). Her parents were *Delavan Ford* (1830/32– 29 Nov 1905; Milwaukee, Wisconsin) and *Sarah Loomis Dewey* (1839–?).

According to census records Nellie was living in Chicago in 1920 and 1930. She died 3 December 1936 in Chicago, Illinois (*Cook County, Illinois Death Index, 1908–1988*). The Deweys can trace their lineage back to several other prominent Puritan families, including the Loomises, Binghams, Phelps, Strongs, and Ingersolls.

Lauchlin A. Mitchell was born in Scotland on 8 May 1855 (according to his death certificate). He was the son of Robert Mitchell and Flora MacKinnon (d. 1876?), both of Scotland. He immigrated to America and married Elizabeth J. Gleason. They lived in the Chicago area. He died in Chicago on 1 February 1914 in a carriage accident. He is buried at Oakwoods Cemetery in Chicago.

Elizabeth J. Gleason was born in Indiana (1930 Federal Census) on 23 June 1861. She was the daughter of Michael Gleason (b. 1822) and Mary Curtin. She died 11 September 1954.

Selected Sources For Genealogical Research

Ancestry.com. 2006. Connecticut town birth records, pre-1870 (Barbour collection) [online database]. Ancestry.com Operations Inc., Provo, UT.

Ancestry.com. 2008. Cook County, Illinois death index, 1908-1988 [online database]. The Generations Network, Inc., Provo, UT.

Anderson RC. 1995. *The great migration begins: Immigrants to New England 1620-1633*, Volume 1. New England Historic Genealogical Society, Boston.

Barbour LB. 2001. *Families of early Hartford, Connecticut.* Genealogical Publishing Co., Inc., Baltimore.

Bowler NP, Malone CCB. 1905. *Record of the descendants of Charles Bowler, England—1740—America, who settled in Newport, Rhode Island.* The Forman-Basset-Hatch Co., Cleveland.

Burrage HS, Stubbs AB. 1909. *Genealogical and family history of the state of Maine.* Lewis Historical Publishing Co., New York.

Cheney TA. 1868. *Historical sketch of the Chemung Valley, etc.* Watkins, NY.

Colonial Dames of America. 1910. *Ancestral records and portraits: A compilation from the archives of Chapter I, the Colonial Dames of America.* The Grafton Press, New York.

Crafts JM, Crafts WF. 1893. *The Crafts family: A genealogical and biographical history of the descendants of Griffin and Alice Craft, of Roxbury, Mass. 1630–1890.* Gazette Printing Company, Northampton, MA.

Crane EB. 1907. *Historic homes and institutions and genealogical and personal memoirs of Worcester County, Massachusetts.* Lewis Historical Publishing Co., New York.

Cutter WR. 1913. *New England families, genealogical and memorial: A record of achievements of her people in the making of the commonwealths and the founding of a nation.* Lewis Historical Publishing Company, New York.

Dedham Historical Society. 1894. *The Dedham historical register*, Volume V. Dedham, MA.

Dewey AM, Dewey LM. 1898. *Life of George Dewey, Rear Admiral, U.S.N.; and Dewey family history.* Dewey Publishing, Westfield, MA.

Dexter FB. 1885. *Biographical sketches of the graduates of Yale college history, 1701–1815.* H. Holt and Company, New York.

Dodd J. 2004. *Illinois marriages, 1790–1860* [online database]. The Generations Network, Inc., Provo, UT.

Dodd J. 2005. *Massachusetts marriages, 1633–1850* [online database]. The Generations Network, Inc., Provo, UT.

Hatcher PL. 2007. *Abstract of graves of revolutionary patriots.* Heritage Books, Westminster, MD.

Illinois General Assembly. 1867. *Private laws of the state of Illinois, passed by the Twenty-Fifth General Assembly, convened January 7, 1867,* Volume II. Baker, Bailhache & Co., Springfield, IL.

Loomis E. 1875. *The descendants of Joseph Loomis, who came from Braintree, England in the year 1638, and settled in Windsor, Conn., in 1639.* Tuttle, New Haven, CT.

Pike County Historical Society, Pittsfield, IL.

Pike Family Association. 1905. *Records of the Pike family association of America.* Press of G.H. & A.L. Nichols, Lynn, MA.

Power JC, Power SA. 1876. *History of the early settlers of Sangamon County, Illinois: Centennial record.* E.A. Wilson & Co., Springfield, IL.

Smith HA. 1889. *A genealogical history of the descendants of the Rev. Nehemiah Smith of New London County, Conn., 1638–1888.* Joel Munsells Sons, Albany, NY.

Stearns ES, Runnels MT. 1906. *History of Plymouth, New Hampshire.* University Press, Cambridge, MA.

Thompson JM. 1967. *The Jess M. Thompson Pike County history: As printed in installments in the* Pike County Republican, *Pittsfield, Illinois, 1935–1939.*

Torrey CA, Bentley EP. 1985. *New England marriages prior to 1700.* Genealogical Publishing Co., Baltimore.

Wheeler RA. 1875. *History of the First Congregational Church, Stonington, Conn., 1674–1874.* T.H. Davis & Company, Norwich, CT.

Whitmore WH, Wyman TB. 1878. *The graveyards of* Boston. First volume, *Copp's Hill epitaphs.* Joel Munsell, Albany, NY.

Sources

The following people and institutions are gratefully acknowledged for providing family information and additional research: Curtis Mann, William Rapai, Dana C. Ewell, John Zarrillo, Stephanie Satalino, Maureen Berejka, Elizabeth Watson, Alex Gann, Matthew Meselson, Chicago History Museum, and the Lincoln Library at the Public Library of Springfield, Illinois.

The CSHL Archives Repository contains a wealth of information and documents from James D. Watson's personal collection. This Repository can be accessed at http://libgallery.cshl.edu. The repository and metadata were created in part through a two-year grant funded collaboration with the Wellcome Library Codebreakers: Makers of Modern Genetics digitization project.

Chapters 1 and 2

1888 Chicago city directory. President of the Concordia Gold Mining Company, 225 Dearborn St., Chicago, residing in Lake Geneva.

1889 Chicago city directory. President of the Concordia Gold Mining Company, 225 Dearborn St., Chicago, residing in Lake Geneva.

49er Quartz Gold Nugget & Unrecorded 49er Constitution. Michael Vinson Americana. http://michaelvinson.com/49er-quartz-gold-nugget-unrecorded-49er-constitution/.

A gold find in Honduras. In *The New York Times*, January 27, 2886.

A scrapbook of Abraham Lincoln's inaugural train trip. *The New York Times*, February 10, 2011. http://www.nytimes.com/interactive/2011/02/10/opinion/disunion_scrap book.html?ref=abrahamlincoln.

Abraham Lincoln Historical Digitization Project: Lincoln/Net. *Lincoln's biography*. Available at http://lincoln.lib.niu.edu/aboutinfo.html.

Advertisement. A fine established building for sale. In *Daily Illinois State Journal* (Springfield, IL), January 23, 1865.

Advertisement. A fine established business for sale. In *Daily Illinois State Register*, January 23, 1865, p. 3.

Advertisement. Bakery and Confectionary. In *Illinois Weekly State Journal* (Springfield, IL), August 5, 1842, p. 2.

Advertisement. Confectionary, &c. In *Illinois Weekly State Journal* (Springfield, IL), February 23, 1839, p. 4.

Advertisement. Cordials, Candies, &c. In *Illinois Weekly State Journal* (Springfield, IL), October 22, 1836, p. 4.

Advertisement. Crackers. In *Illinois Weekly State Journal* (Springfield, IL), August 20,1841, p. 2.

Advertisement. Groceries, Confectionary, &c. In *Illinois Weekly State Journal* (Springfield, IL), January 27, 1848, p. 3.

Advertisement. Kossuth, etc. *Daily Illinois State Register*, April 21, 1852, p. 2.

Advertisement. Mayoralty of Springfield. In *Illinois Weekly State Journal* (Springfield, IL), April 2, 1846, p. 3.

Advertisement. Notice. *Daily Illinois State Register*, October 29, 1860, p. 4.

Advertisement. Water for the Million! In *Daily Illinois State Journal*, August 15, 1876.

Advertisement. Watson's Cough Candy. In *Illinois Weekly State Journal* (Springfield, IL), January 18, 1844, p. 3.

Advertisement. Watson's Restaurant Francais, 145 Dearborn Street, Chicago. *Illinois State Journal*, January 23, 1882, p.3

Advertisement. W.W. Watson, 202 LaSalle St., Real Estate Broker and Dealer in Chicago and Cook County, Washington Heights. *Daily Illinois State Journal*, May 4, 1874.

Advertisement. W.W. Watson & Son. In *Daily Illinois State Register*, January 14, 1856, p. 3.

Advertisement. W.W. Watson having taken his son B.A. Watson into co-partnership. *Illinois Weekly State Journal* (Springfield, IL), November 8, 1839.

Advertisement. W.W. Watson. In *Illinois Weekly State Journal* (Springfield, IL), February 23, 1839, p. 1.

All Chicago knows him. In *Daily Illinois State Register,* March 25, 1896, p. 9.

Angle PM. 1935. Here I have lived. In *A history of Lincoln's Springfield* (1821–1865), pp. 252–255. The Abraham Lincoln Association, Springfield, IL.

Atlas map of Pike County, Illinois. 1872. Compiled, drawn, and published from personal examinations and surverys by Andreas, Lyter & Co. Reprinted by Unigraphic, Evansville, IN, 1976.

Atlas Map of Pike County, Illinois/compiled, drawn, and published from personal examinations and surveys by Andreas, Lyter & Co., 1872. Reprinted by the Pike County Historical Society, Pittsfield, IL, 1976.

Bartow E, Udden JA, Parr SW, Palmer GT. 1908. *The mineral content of Illinois waters*. Water Survey Series No. 4, University of Illinois Bulletin, Vol. 6, no. 3, p. 67. University of Illinois, Urbana, IL.

Brown CO. 1922. Springfield society before the Civil War. *Journal of the Illinois State Historical Society* 15: 477–500.

Bulliet CJ. 1970. Artists of Chicago past and present: Dudley Crafts Watson. *Chicago Tribune*, March 8, 1970.

Burlingame M. 2013 [reprint edition]. *Abraham Lincoln: A life*, Vol. 1, Ch. 7. Johns Hopkins University Press, Baltimore.

By this morning's mails. In *Daily Illinois State Journal*, June 20, 1850, p 2.

Campbell & Richardson's Springfield City Directory and Business Mirror for 1863, pp. 84 and 134. Johnson & Bardford, Booksellers, and Stationers, Springfield, IL.

Chicago and Northwestern Railway Company. 1882. *My rambles in the enchanted summer land of the Great Northwest during the tourist season of 1881*, pp. 5–6. Rand, McNally, Chicago.

Chicago city directories 1872–1876, 1878.

Constitution of the Illlinois and California Mutual Insurance Company No. 1. Abraham Lincoln Presidential Library & Museum, Springfield, IL.

Dodd J. 1997. Sarah Mottashed–William W. Watson. In *Illinois marriages to 1850*. Provo, UT. Ancestry.com Operations, Inc.

Drury J. 1977. *Old Illinois houses*. Reprinted by The University of Chicago Press, Chicago. Available at http://penelope.uchicago.edu/Thayer/E/Gazetteer/Places/America/United_States/Illinois/_Texts/DRUOIH/Central_Illinois/29*.html.

Dubois JK. JK Dubois to Abraham Lincoln, Saturday, May 23, 1863 (Complaints about Ninian Edwards and William Bailhache; endorsed by William Yates, et al.)

Abraham Lincoln Papers at the Library of Congress. Transcribed and Annotated by the Lincoln Studies Center, Knox College, Galesburg, IL.

Dubois JK, Hatch OM, et al. to Abraham Lincoln, Thursday, June 4, 1863 (appointment of commissary and quartermaster; endorsed by Lincoln, June 22, 1863). The Abraham Lincoln Papers at the Library of Congress. Transcribed and Annotated by the Lincoln Studies Center, Knox College, Galesburg, IL.

E.H. Hall's Springfield City Directory and Sangamon County Advertisers for 1855–1856.

Ewell Family History. Privately published. Courtesy of Dana C. Ewell.

Farnsworth KB, Walthall JA. 2011. *Bottled in Illinois: Embossed bottles and bottled products of early Illinois merchants from Chicago to Cairo, 1840–1880*. Illinois State Archaelogical Survey, Urbana, IL.

Filege S. 2002. *Tales and trails of Illinois*, p. 105. University of Illinois Press, Champaign, IL.

Find A Grave Index, 1800–2012, Provo, UT. http://www.findagrave.com.

Fine lectures free. In *Chicago Daily News*, March 12, 1913, p. 13.

Fire at Lake Geneva. *Kalamazoo Gazette* (Kalamazoo, MI), July 10, 1894, p. 7.

For the Illinois State Register. In *the Illinois State Register*, March 12, 1851, p. 2.

Former Springfield man dies in Chicago [W.W. Watson III obituary]. In *Daily Illinois State Register*, March 12, 1913, p. 3.

From California. In *Daily Illinois State Journal*, April 20, 1850.

Gay travelers—Events (Reception for the Vice-President of Honduras). In *Daily Inter Ocean* (Chicago, IL), August 14, 1887, pt. 2, p. 16.

Guyape Gold Inc. *Daily Inter Ocean* (Chicago, IL), May 17, 1888, pt. 1, p. 8.

Hadley AT. 1921. The baccalaureate sermon. *Yale Alumni Weekly* 30: 1115, 1116.

Hart RE. *Lincoln's Springfield: The Public Square (1823–1865)*. Available at http://lincolns-springfield.blogspot.com/.

History of Pike County, Illinois: Together with sketches of its cities, villages and townships, (etc.), pp. 474, 475, 511, 906. 1880. Chas. C. Chapman & Co, Chicago.

History of Sangamon County, Illinois, pp. 753, 754. 1881. Inter-State Publishing Company, Chicago.

History of Walworth County, Wisconsin, p. 881. 1882. Western Historical Company, Chicago.

Hoch BR. 2001. *The Lincoln Trail in Pennsylvania: A history and guide*. Penn State University Press, University Park, PA.

Holzer H. 2008. *Lincoln President-Elect: Abraham Lincoln and the Great Secession, Winter 1860–1861*, pp. 38–45. Simon & Schuster, New York.

Honduras Mine. *Daily Inter Ocean* (Chicago), October 7, 1888, pt. 2 pp. 10–11.

How The Result Is Received; At The West. From the Home of Mr. Lincoln How He Received the News Speculation as to His Course, &c. Special Dispatch to the *New-York Times*, November 8, 1860. http://www.nytimes.com/1860/11/08/news/result-received-west-home-mr-lincoln-he-received-speculation-his-course-c.html.

Ives, Mrs. John G. (Abigail Watson) made a flag for Pres. Abraham Lincoln, 15 June 1915, Springfield, Sangamon, Illinois, USA. *Journal of the Illinois State Historical Society* VIII [April 1915 to January 1916]: 344–346.

James D. Watson pedigree. Available at watson.ancestortree.net.

Kendeigh CS. 1952. In Memoriam: Lynds Jones. *Auk* 69: 258–265.

LaVictoria Mining Inc. *Illinois State Journal,* July 1, 1887, p. 6.

Lincoln A. Abraham Lincoln. Lincoln check to Watson & Son. *via*Libri. http://www.vialibri.net/552display_i/year_1855_0_96973.html.

Lincoln A. Abraham Lincoln to Jesse K. Dubois et al., Friday May 29, 1863. The Abraham Lincoln Papers at the Library of Congress. Series 1. General Correspondence, 1833–1916.

Lincoln A. 1953. *Collected works of Abraham Lincoln, 1809–1865*, Vol. 7. Rutgers University Press, New Brunswick, NJ.

Lincoln Home. National Park Service. Available at http://www.nps.gov/liho/historyculture/index.htm.

Lincolniana Notes. 1954. Recollections of a Springfield doctor (Dr. Preston H. Bailhache). *Journal of the Illinois State Historical Society* 47: 57–63.

Maine Birth Records 1621–1922. Augusta Crafts Tolman. City Clerk's Office, Portland, ME.

Menz KB. 1983. Historic Furnishings Report. Lincoln Home. National Historic Site/Illinois, Sections A through E. The Lincoln Home. Lincoln Home National Historic Site, Springfield, IL. Harpers Ferry Center, National Park Service, U.S. Department of the Interior, pp. 19–20.

Mr. B.A. Watson, from California…. In *Daily Illinois State Journal*, January 30, 1851.

Nellie Dewey Watson [Nellie Dewey Ford]. Illinois, Deaths and Stillbirths Index, 1916–1947. Provo, UT. FHL Film Number 1927308.

Obituary. B.A. Watson. In *Daily Illinois State Register*, Sunday, June 23, 1901, p. 5, col. 3.

Obituary. John G. Ives. Batavia, Illinois. *Batavia [Illinois] Herald*, April 7, 1898.

Obituary. W.W. Watson. In Daily Illinois State Register, Wednesday, March 12, 1913, page 3, col. 4.

Obituary of the late William W. Watson, Springfield, Illinois. November 2, 1874. Source unknown.

Pike County Historical Society Museum, Pittsfield, IL.

Pinsker M. 2008. Not always such a Whig: Abraham Lincoln's partisan realignment in the 1850s. *Journal of the Abraham Lincoln Association* 29: 27–43. Available at http://quod.lib.umich.edu/j/jala/2629860.0029.204/—not-always-such-a-whig-abraham-lincolns-partisan-realignment?rgn=main;view=fulltext.

Power JC. 1876. History of the early settlers of Sangamon County, Illinois. *Centennial Record*, pp. 480, 754. Edwin A Wilson & Co., Springfield, IL.

Pratt HE (compiled by). 1944. *Concerning Mr. Lincoln: In which Abraham Lincoln is pictured as he appeared to letter writers of his time.* The Abraham Lincoln Association, Springfield, IL.

Private Laws of the State of Illinois passed by the twenty-fifth General Assembly, Vol. I. 1867. Baker, Bailhache & Co, Springfield, IL.

Rawls JJ, Orsi RJ, eds. 1999. *A golden state: Mining and economic development in Gold Rush California.* University of California Press, Berkeley, CA. http://ark.cdlib.org/ark:/13030/ft758007r3/.

Reception for the vice president of Honduras. *Daily Inter Ocean* (Chicago), August 14, 1887, pt. 2, p. 16.

Ruger A. 1867. Springfield, Illinois 1867. In *Drawn from nature by A. Ruger.* Chicago Lithographing Co., Chicago.

Sangamon County Historical Society, Sangamon Valley Collection, Lincoln Library, Springfield, IL.

Springfield Board of Trade. 1871. *History of Springfield, its attractions as a home and advantage for business and manufacturing, etc.*, pp. 15, 41.

The Illinois and California Mining Mutual Insurance Company. In *Daily Illinois State Journal*, March 27, 1849, p. 2.

Thomas Tolman Watson. Draft registration card, 3619, 4427. Dated December 1, 1916.

Thomas Tolman Watson. Illinois, Deaths and Stillbirths Index, 1916–1947. Provo, UT. FHL Film Number 1892488.

To check some misapprehensions that seem to be in circulation…. *Daily Illinois State Registe*r, March 23, 1850, p. 2.

Topics in chronicling America—St. Louis World's Fair (The Louisiana Purchase Exposition, 1904). The Library of Congress Newspaper & Current Periodical Reading Room. Available at http://www.loc.gov/rr/news/topics/stlouis.html.

Unveiling of Lincoln Marker by Springfield Chapter, Daughters of The American Revolution—14 June 1915. *Journal of the Illinois State Historical Society* VIII [April 1915 to January 1916]: 344–346.

U.S. Census. 1850. Sangamon County, Springfield Illinois, p. 123.

U.S. Census. 1860. District no. 16, Springfield Illinois, p. 126.

U.S. Census. 1880. Village of Geneva, Walworth County, Wisconsin. Enumeration district 227, p. 14.

U.S. Census. 1900. *Twelfth census of the United States, 1900.* Hyde Park, Cook County, Illinois, Enumeration District 1014, Sheet no. 3.

U.S. Census. 1900. *Twelfth census of the United States, 1900.* Fayal Township, St. Louis county, Minnesota, Enumeration district no. [illeg], sheet no. 4.

U.S. Census. 1910. *Thirteenth census of the United States, 1910.* Tract G14, Cook County, Illinois. Enumeration district no. 403, sheet no. 18.

U.S. Census. 1920. *Fourteenth census of the United States, 1920.* Chicago City, Cook County, Illinois. Enumeration district no. 381, ward 7, sheet no. 6.

U.S. City Directories, 1821–1989 database. Provo, UT. 1863, 1864.

Victoria Mining Excursion. *Daily Inter Ocean* (Chicago), August 7, 1887, pt. 2, p. 16.

Watson BA. 1849–1851. Benjamin A. Watson Gold Rush letters, 1849–1851. Manuscript 3153; no online items. Online Archive of California, University of California.

Watson JD Sr. World War I letters. Cold Spring Harbor Laboratory Archives, Cold Spring Harbor, NY.

Webb v. Watson. 1837. In *Law practice of Abraham Lincoln*, 2nd ed., 2006, 2008. The Papers of Abraham Lincoln. Available at www.lawpracticeofabrahamlincoln.org.

Western Historical Co. 1882. *History of Walworth County, Wisconsin*, p. 881. Western Historical Company, Chicago.

Williams CS. 1860. *Williams' Springfield directory: City guide and business mirror for 1860–1861*, pp. 95, 103, 140. Johnson & Bradford Booksellers and Stationers, Springfield, IL.

Chapter 3

Aretino P. 1534–1536. *Dei Ragionamenti*. L'Editrice del Libro Rare, Milan [1920 reprint].

Baatz S. 2008. *For the thrill of it: Leopold, Loeb, and the murder that shocked Chicago.* HarperCollins, New York.

Baatz S. 2008. Leopold and Loeb's criminal minds. *Smithsonian Magazine*, August 2008. Available at http://www.smithsonianmag.com/history-archaeology/criminal-minds.html?c=y&page=4.

Berea College. William J. Hutchins. http://community.berea.edu/hutchinslibrary/specialcollections/bca/rg03.asp.

Business & Finance: Chicago. *Time,* June 15, 1931.

Convict kills Loeb, Franks boy slayer. *The New York Times,* January 29, 1936.

Darrow C. 1922. *Crime: Its cause and treatment.* Thomas Y. Crowell Company, New York.

Darrow C. 1924. *Attorney Clarence Darrow's plea for mercy and Prosecutor Robert E. Crowe's demand for the death penalty in the Loeb-Leopold case, the crime of the century.* Wilson Publishing Company, Chicago.

Darrow C. 1932. *Story of my life.* Charles Scriber's Sons, New York.

Darrow C. Closing Argument. The State of Illinois v. Nathan Leopold & Richard Loeb. Delivered by Clarence Darrow. Chicago, Illinois, August 22, 1924. Available at http://law2.umkc.edu/faculty/projects/ftrials/leoploeb/darrowclosing.htm.

Dendroica kirtlandii. Birds Collection Database. Field Museum of Natural History, Bird and Egg Collections. PMNH number 60981, June 19, 1923.

Dzuback MA. 1991. *Robert M. Hutchins: Portrait of an educator.* University of Chicago Press, Chicago.

Eggers Wood Forest Preserve. 2005. Chicago Habitat Directory, City of Chicago, p. 54. Available at http://www.cityofchicago.org/city/en/depts/dcd/supp_info/nature_areas_directory.html.

Franks death letter like current story in magazine; Franks inquest awaits report from chemist; Queries race theories based on boy murder; Kidnapers' ransom letter shows hand of expert letterer. All available at American Newspaper Repository, Duke University Libraries.

Expert fixes on kind of machine kidnapper used. *Chicago Daily Tribune*, May 24, 1924. Available at Homicide in Chicago 1870–1930 at http://homicide.north western.edu/ crimes/leopold/newspaper/19240524trib04/.

Groves A. 2006. *World's Columbian Exposition fire, Chicago: July 10, 1893*. Available at https://www.ideals.illinois.edu/handle/2142/84.

Guide to Harold S. Hulbert papers, Northwestern University Archives, Evanston, IL. Available at http://findingaids.library.northwestern.edu/catalog/inu-ead-nua-archon-167.

Guide to the Bernard Glueck, Sr. papers, 1910–1971. University of Chicago Library, Chicago, IL. Available at http://www.lib.uchicago.edu/e/scrc/findingaids/view.php?eadid=ICU.SPCL.GLUECKB.

Guide to the Leopold and Loeb Collection. Northwestern University, Evanston, IL. Available at http://findingaids.library.northwestern.edu/catalog/inu-ead-spec-archon-1458.

Hennigar MJ. 1986. The first Paul Bunyan story in print. *Journal of Forest History* October 1986: 175–177.

Higdon H. 1994. *Leopold and Leob: The crime of the century*. University of Illinois Press, Champaign, IL.

His Chicago DNA: Genetic pioneer James Watson recalls a South Side education. *Chicago Tribune*, January 30, 2004. http://articles.chicagotribune.com/2004-01-30/features/0401300053_1_james-d-watson-exhibit-father.

History. Jackson Park. Available at http://www.chicagoparkdistrict.com/parks/jackson-park/.

Hoffman D. 1986. The birth of Paul Bunyan—In print. *Journal of Forest History* October 1986: 177–178.

Houston K. 1996. The man who could out-lumber Paul Bunyan. *The Detroit News*, June 14, 1996. Available at http://apps.detnews.com/apps/history/index.php?id=154.

Indictment, Nathan F. Leopold and Richard Loeb. 1924. Criminal Court of Cook County, State of Illinois.

Jackson Park History. Chicago Park District. Available at www.chicagopark district .com/parks/jackson-park.

Jackson Park's Wooded Island. Jackson Park Advisory Council and the Chicago Park District. Available at http://jacksonparkadvisorycouncil.org/the-wooded-island .html.

Joked with girl about the murder. *The New York Times*, June 2, 1924.

Kirtland's Warbler (*Dendroica kirtlandii*) species profile. Environmental Conservation Online System, FWS Endangered. Available at http://ecos.fws.gov/speciesProfile/ profile/speciesProfile.action?spcode=B03I.

Leopold and Loeb Collection. Northwestern University. Available at http://findingaids. library.northwestern.edu/catalog/inu-ead-spec-archon-1458.

Leopold and Loeb Trial. The Clarence Darrow Digital Collection. University of Minnesota Law Library. Available at http://darrow.law.umn.edu/trials.php?.

Leopold faces operation in prison. *The New York Times*, May 26, 1925.

Leopold NF Jr. 1923. Reason and instinct in bird migration. *Auk* XL: 409–414.

Leopold NF Jr. 1924. The Kirtland's Warbler in its summer home. *Auk* XLI: 44–58.

Leopold NF. 1958. *Life plus 99 years*. Doubleday, New York.

Leopold NF. 1963. *Checklist of birds of Puerto Rico and the Virgin Islands*. University of Puerto Rico, Agricultural Experiment Station, Rio Piedras, Puerto Rico.

Leopold received parole in 1924 "thrill" slaying. *The New York Times*, February 21, 1958.

Leopold spends second freedom day in Chicago. *The News Tribune*, March 14, 1958, p. 12. Available at http://www.newspapers.com/newspage/31978194/; also *Miami Daily News-Record*, March 14, 1958, p. 1. Available at http://www.newspapers .com/newspage/5568138/.

Linder DO. 1997. *The Leopold and Loeb trial: A brief account.* University of Missouri–Kansas City School of Law. Available at http://law2.umkc.edu/faculty/ projects/ftrials/leoploeb/accountoftrial.html.

Linder DO. *Nathan Leopold and ornithology*. University of Missouri–Kansas City School of Law. Available at http://law2.umkc.edu/faculty/projects/ftrials/ leoploeb/LEO_ORNI.htm.

Linder DO. *Richard Loeb (1905–1936)*. University of Missouri–Kansas City School of Law. Available at http://law2.umkc.edu/faculty/projects/leopleob/ LEO_LOEB.htm.

Linder DO. *Tennesse vs. John Scopes: The "Monkey Trial."* University of Missouri–Kansas City. http://law2.umkc.edu/faculty/projects/ftrials/scopes/scopes.htm.

Linder DO. *The glasses: The key link to Leopold and Loeb.* University of Missouri–Kansas City School of Law. Available at law2.umkc.edu/faculty/proj ects/ftrials/leoploeb/LEO_GLAS.htm.

Loeb leader, real slayer, alienists say. *Chicago Tribune.* Available at http://homicide.northwestern.edu/docs_fk/homicide/5866/19240722trib01.jpg.

Loss on exhibits not heavy: Damage by World's Fair fire largely overestimated. *The New York Times*, January 10, 1899.

MacGillivray J. *The Kirkland warbler in its home: Third known nest* [motion picture]. Chicago History Museum, Chicago, IL. Call no.: 1982.0020 F0001.

Mayer M. 1993. *Robert Maynard Hutchins: A memoir.* University of California Press, Berkeley. Available at http://ark.cdlib.org/ark:/13030/ft4w10061d/.

Memorial. Norman Asa Wood. Faculty History Project, University of Michigan. Available at http://um2017.org/faculty-history/faculty/norman-asa-wood/memorial.

Mullen W. 2004. His Chicago DNA: Genetic pioneer James Watson recalls a South Side education. *Chicago Tribune*, January 30, 2004. http://articles.chicagotribune.com/2004-01-30/features/0401300053_1_james-d-watson-exhibit-father.

Nathan F. Leopold, Jr. Passport no 402388. Issued April 28, 1924. Department of State, United States.

Nathan Leopold allowed to wed. *The New York Times*, January 20, 1961.

Palmer TS. 1924. The forty-first stated meeting of the American Ornithologists' Union. *Auk* XLI: 122–123.

Penitentiary Mittimus, Nathan F. Leopold, Junior. Criminal Court of Cook County, Chicago, IL.

Pittman V. 2009. *Correspondence study and the "crime of the century": Helen Williams, Nathan Leopold, and the Stateville Correspondence School*, Vol. 26. Vitae Scholasticae Publisher: Caddo Gap Press, San Francisco. Available at http://www.freepatentsonline.com/article/Vitae-Scholasticae/277602571.html.

Rapai W. 2012. *Kirtland's Warbler: The story of a bird's fight against extinction and the people who saved it.* University of Michigan Press, Ann Arbor, MI.

Rydell RW. 2005. World's Columbian Exposition (May 1, 1893–October 30, 1893). *Electronic encyclopedia of Chicago.* Chicago Historical Society and The Newberry Library, Chicago, Illinois. Available at http://www.encyclo pedia.chicago history.org/pages/1386.html.

Spring Migration Notes of the Chicago Area—Review. 1920. *Auk* 37: 616.

Statement of Nathan F. Leopold, Jr. State's Attorney of Cook County, Chicago. http://homicide.northwestern.edu/crimes/leopold/.

Statement of Richard Albert Loeb. State's Attorney of Cook County, Chicago. http://homicide.northwestern.edu/crimes/leopold/.

The Crime of the Century. People v. Nathan F. Leopold, Jr., and Richard Loeb, Criminal Court #33623 and 33624. Records and Archives, Clerk of the Circuit Court, Cook County, Illinois. The file contains the indictments, subpoenas (grand jury and witness), judge's warnings to the defendants, various capias and appearances, and the prison mittimus. The Archives also holds a photocopy of the 11-volume file transcript of *People v. Leopold and Loeb, #33623 and 33624*, generously provided by Northwestern University Library's Special Collections (www.library.nwu.edu/spec/). Available at http://www.cook countyclerkofcourt.org/?section=RecArchivePage&RecArchivePage=leo pold_and_loeb.

The World's Columbian Exposition. Chicago Historical Society. Available at www .chicagohs.org/history/expo.html.

Thomson T. *Birds of a feather: Nathan Leopold's Warbler*. Available at www.net walk.com/~vireo/021.html.

Two students held in Franks slaying. *The New York Times*, May 31, 1924.

Watson JD, Lewis GP, Leopold NF Jr. 1920. *Spring migration notes of the Chicago area*. Privately printed, 18 pp.

Wiley JW. 1996. Ornithology in Puerto Rico and the Virgin Islands. *Ann NY Acad Sci* 776: 149–179.

World's Columbian Exposition. 2005. *Electronic encyclopedia of Chicago*. Chicago Historical Society, The Newberry Library. Available at www.encyclopedia.chi cagohistory.org/pages/1386.html.

Chapter 4

14-year-old girl inherits $10,000,000. J.H. Barker, car builder, leaves bulk of his estate to his daughter. *The New York Times*, December 6, 1910.

Alfonso Iannelli–Dudley Crafts Watson Correspondence. Kelmscott Gallery. Available at http://www.ebay.com/itm/Alfonso-Iannelli-Dudley-Crafts-Watson-Correspondence-1924-/360652418394.

Art Institute of Chicago. Museum Education at the Art Institute of Chicago, Chicago, IL.

A Sunday on La Grande Jatte. Art Institute of Chicago, Chicago. Available at http://www.artic.edu/aic/collections/artwork/27992.

Bernard Ewell Art Appraisals, LLC, Santa Fe, New Mexico 87508.

Better Business Letters. 1920. *The Rotarian* XVI(4): 177.

Callow S. 1997. *Orson Welles: Volume 1: The road to Xanadu*. Penguin Books, New York [reprint edition].

Death of Michael Gleason. Newspaper clipping, source unknown.

Dudley Crafts Watson: Obituary. *Milwaukee Journal*, December 26, 1972.

Ewell B. Phone conversation, May 28, 2013.

Ewell Family Heritage. Privately published. Dana C. Ewell, Cincinnati, OH 45244.

Higham C. 1985. *Orson Welles: The rise and fall of an American genius*. St. Martin's Press, New York.

Hill R, Welles O. 1938. On the teaching of Shakespeare and other great literature. *The English Journal*, June 1938. Available at http://www.wellesnet.com/?p=269.

Historic Structures of Michigan City, Indiana. Available at http://www.preserve indiana.com/pixpages/nw_ind/michcity.html.

James A. 2013. *LaSalle Extension University, snail-mail generations' University of Phoenix*. Available at http://boingboing.net/2013/02/lasalle-extension-un.html.

Jesse G. Chapline, head of LaSalle Extension University, dies in Chicago. *The New York Times*, July 6, 1937.

John H. Barker. LaPorte County, Indiana Index to Marriage Records, Letters A–C, volume I1. Compiled by Indiana Works Progress Administration, Book G, p 293, 1938–1940.

MacLiammoir M. 1946. *All for Hecuba: An Irish theatrical autobiography*. Methuen, London.

Mitchell LA. Certification of Record of Death (February 1, 1914), February 9, 1914, no. 12192. Department of Health, City of Chicago, Chicago.

Mitchell, Lauchlin A. Cook County, Illinois, Deaths Index, 1879–1922. Illinois Department of Public Health, Division of Vital Records, Springfield, IL. FHL film no. 1239969.

Mitchell MJ. Margaret Jean Mitchell. CSHL Archives Repository. Available at http://libgallery.cshl.edu/items/show/51115 and http://libgallery.cshl.edu/items/show/511159.

Morris HC. 1902. John H. Barker. In *The history of the First National Bank of Chicago,* p. 173. R.R. Donnelley & Sons Company, Chicago.

Oliver M. 2002. Orson and Rita. In *Mike Oliver's Acapulco*, pp. 165–166. iUniverse.com, pp. 165–166.

Olvaney cousins. CSHL Archives Repository. http://libgallery.cshl.edu/items/show/51116.

Packard J. 1876. Coolspring Township. In *History of La Porte Count, Indiana*, Chapter XIII. Available at http://www.dunelady.com/laporte/all_histories.htm.

Radio listeners in panic, taking war drama as fact. *The New York Times*, October 31, 1938.

Richard Head Welles. U.S. World War I Draft Registration Cards, 1917–1918. U.S. Selective Service System, National Archives and Records Administration, Washington, DC, M1509, FHL roll no. 1503828.

Richard H. Welles. 1928. New York Passenger Lists, 1820–1957. S.S. *Fort St. George*, no. 6. Provo, UT.

Schnitzer V. U-M Now Home to World's Most Extensive Orson Welles Archive. *Michigan Today*, October 23, 2012. http://michigantoday.umich.edu/story.php?id=8480#.UrDDpva8UUg.

Shook SR. Steamer "Roosevelt," 1912—Michigan City Indiana. Postcard Brink Publishing Company. Available at www.flickr.com/photos/shookphotos/6368856623.

Stevenson to Quit Law. Former Vice President will aid LaSalle Extension University. *The New York Times*, March 2, 1909.

Styles JHT. Roger Hill's daughter recalls Orson Welles at the Todd School for Boys in Woodstock, Illinois. Orson Welles Web Resource. http://www.wellesnet.com/?p=5005.

Tarbox T. 2013. *Orson Welles and Roger Hill: A friendship in three acts*. BearManor Media.

The Barker Mansion. Healthy Communities of LaPorte County, LaPorte IN.

The Orson Welles Archive at the University of Michigan. http://quod.lib.umich.edu/cgi/f/findaid/findaid-idx?c=sclead&idno=umich-scl-wilsonwelles.

The Spanish Civil War. History World International. Available at http://history-world.org/spanish_civil_war.htm.

U.S. Census. 1930. *Fifteenth census of the United States, 1930*. Chicago City, Cook Country, Enumeration district no. 16-304, sheet 7-A.

U.S. Census. 1940. *Sixteenth census of the United States, 1940*. Chicago City, Cook County, Illinois. S.D. no. 26, sheet no. 6-A.

U.S. Census. 1990. *Twenty-first census of the United States, 1990*. Chicago, County, IL. Ross 285, page 3A, enumeration district no. 0993.

Walkinshaw LH. 1983. *Kirtland's Warbler: The natural history of an endangered species*. Cranbrook Institute of Science, Bloomfield Hills, MI.

Watson JD Sr. 1939–1940. Diary Extracts. Cold Spring Harbor Laboratory Archives, Cold Spring Harbor, NY.

Welles O. 1982–1983. My father wore black spats, and A brief career: As a musical prodigy. *Paris Vogue*, December 1982/January 1983. Available at http://www.wellesnet.com/phpbb2/viewtopic.php?f=2&t=748.

Welles O. Sketchbook Episode 5. Orson Welles Web Resource. Available at http://www.wellesnet.com/sketchbook5.htm.

Wendell Phillips Academy High School. History at http://www.phillips.cps.k12.il.us/history.html.

Chapter 5

About Enrico Fermi. From Biographical note, Guide to the Enrico Fermi Collection, Special Collections, Research Center, University of Chicago Library, Chicago, IL. Available at http://fermi.lib.uchicago.edu/fermibiog.htm.

Adler MJ. 2004. *C250 celebrates, Columbia University, New York*. Available at http://c250.columbia.edu/c250_celebrates/remarkable_columbians/mortimer_j_adler.

Adler MJ. *The blessings of good fortune*. Great Books Academy. Accessed at http://www.greatbooksacademy.org/newsroom/mortimer-j-adler.

Armand Deutsch, 1913–2005. Obituary. *The Chicago Tribune*, August 18, 2005.

Arthur H. Compton—Biographical. 1927. The Nobel Foundation. Available at http://www.nobelprize.org/nobel_prizes/physics/laureates/1927/comptonbio.html.

Arthur H. Compton. Atomic Heritage Foundation, Washington, DC. Available at http://www.atomicheritage.org/index.php/veterans/246.html?task=view.

Arthur Holly Compton, 1892–1962. American Institute of Physics. Available at http://www.aip.org/history/gap/compton/compton.html.

Benton, William (1900–1973). *Biographical directory of the United States Congress*. Available at http://bioguide.congress.gov/scripts/biodisplay.pl?index=B00 0399.

Bruckner DJR. 1962. The day the nuclear age was born. *The New York Times*, November 30, 1962.

Cape Cod Vacation. CSHL Archives Repository. http://libgallery.cshl.edu/items/show/51134.

Chicago Sun. December 4, 1941, Vol. 1, no. 1, p. 1.

Cockcroft J. *The early days of the Canadian and British atomic energy projects*, pp. 18–20. International Atomic Energy Agency, Vienna. Available at www.iaea.org/Publications/Magazines/Bulletin/.../04004701820su.pdf. Access at http://www.iaea.org/Publications/.

Compton AH. 1923. A quantum theory of the scattering of X rays by light elements. *Phys Rev* 21: 483–502.

Edwards G. *Canada's role in the atomic bomb programs of the United States, Britain, France and India—A chronology*. Canadian Coalition for Nuclear Responsibility, Hampstead, QC. Available at http://www.ccnr.org/chronology.html.

Enrico Fermi—biographical. 1938. The Nobel Foundation. Available at http://www.nobelprize.org/nobel_prizes/physics/laureates/1938/fermi-bio.html.

Expert on controversy: Robert Maynard Hutchins. *The New York Times*, January 20, 1967.

Federation of American Scientists. 2008. *FAS history*. Federation of American Scientists, Washington, DC. Available at http://www.fas.org/member/member_history.html.

Firestein M. *William Rainey Harper: Young man in a hurry*. Harper College Library, Palatine, IL. http://dept.harpercollege.edu/library/archives/william raineyharper.html.

Foe of complacency: Robert Maynard Hutchins. *The New York Times*, April 17, 1961.

Goodspeed JW. 1928. *William Rainey Harper*. University of Chicago Press, Chicago.

Grimes W. 2001. Mortimer Adler, 98, dies: Helped create study of classics. *The New York Times*, June 29, 2001.

History. The University of Chicago, Chicago, IL. Available at http://www.uchicago.edu/about/history.

Hutchins RM. 1936. *The higher learning in America*. Yale University Press, New Haven, CT.

Hutchins RM. 1954. *Great Books: The foundation of a liberal education*. Simon & Schuster, New York.

Hutchins RM. Obituary. Robert M. Hutchins, long a leader in educational change, dies at 78. *The New York Times*, May 16, 1977.

Indiana Dunes. Available at http://www.in.gov/dnr/parklake/2980.htm.

Indiana Dunes State Park, Interpretive Master Plan, 2008. Indiana Department of Natural Resources. Available at www.in.gov/dnr/parklake/files/sp-Indiana_DunesIMP_2008.pdf.

Julius Rosenwald (1862–1932). Sears Archives. Available at http://www.searsarchives.com/people/juliusrosenwald.htm.

Lanouette W (with Szilard B). 1992. *Genius in the shadows: A biography of Leo Szilard*. University of Chicago Press, Chicago.

Leo Szilard. In *Staff biographies*. Los Alamos National Laboratory, Los Alamos, NM. Available at http://www.lanl.gov/history/people/L–Szilard.shtml.

Leo Szilard (1898–1964). National Science Digital Library. Available at http://www.atomicarchive.com/Bios/Szilard.shtml.

Leo Szilard dies; A-bomb physicist. *The New York Times*, May 31, 1964.

Mayer M. 1993. *Robert Maynard Hutchins: A memoir*. University of California Press, Berkeley, CA.

Metallurgical Laboratory at the University of Chicago. Department of Energy, Office of Management. Available at http://energy.gov/management/metallurgical-laboratory-university-chicago.

Metallurgical Laboratory. 2000–2004. In *Manhattan Project history*. The Manhattan Project Heritage Preservation Association, Inc. Available at http://www.mphpa.org/classic/HISTORY/H-05f-1.htm.

Myers EW. Letter from Elizabeth Watson Myers to James D. Watson. CSHL Archives Repository. http://libgallery.cshl.edu/items/show/45849.

Notes on James D. Watson, Sr. CSHL Archives Repository, Cold Spring Harbor, NY. Available at http://libgallery/cshl.edu/items/show/51863.

Oliver M. 2001. Orson and Rita. In *Mike Oliver's Acapulco*, p. 166. iUniverse.com.

Patent is issued on first reactor. *The New York Times*, May 19, 1955.

Purdum TS. 2005. Armand Deutsch, Hollywood fixture, dies at 92. *The New York Times*, August 18, 2005.

Robert M. Hutchins weds his assistant. *The New York Times*, May 11, 1949.

Robert M. Hutchins, long a leader in educational changes, dies at 78. *The New York Times*, May 16, 1977.

Robert Maynard Hutchins, 1929–1951. Office of the President, University of Chicago. Available at http://president.uchicago.edu/directory/robert-maynard-hutchins.

Robert Maynard Hutchins, 1899–1977. *The Presidents of the University of Chicago: A centennial view*. Available at http://www.lib.uchicago.edu/e/spcl/centcat/pres/presch05_01.html.

Rosenwald J. 1929. Principles of public giving. *The Atlantic Monthly*, May 1929.

The Chain Reaction: December 2, 1942 and After. An Exhibition in the Department of Special Collections, University of Chicago Library, October 1, 1992–December 4, 1992. Available at http://www.lib.uchicago.edu/e/scrc/exhibits/chainreaction.html.

Van Doren J. 2001. *The beginnings of the Great Books movement at Columbia*. Living Legacies, Columbia University, New York. http://www.columbia.edu/cu/alumni/Magazine/Winter2001/greatBooks.html.

Watson, Margaret Jean. Letter to James D. Watson from Garden House, Hotel Cambridge, Cambridge, United Kingdom, January 18, 1952. CSHL Archives, Cold Spring Harbor, NY.

Watson's Parent's Home. CSHL Archives Repository. Available at http://libgallery.cshl.edu/items/show/51144.

Whitman A. 1973. William Benton dies here at 72; leader in politics and education. *The New York Times*, March 19, 1973.

William Benton—An appraisal. *The New York Times*, March 25, 1973.

William Burnett Benton (1900–1973). U.S. Department of State, Office of the Historian. Available at http://history.state.gov/departmenthistory/people/benton-william-burnett.

William Rainey Harper (1856–1906). Psychology History, Muskingum College. Available at www.muskingum.edu/~psych/psycweb/history/harper.htm.

William Rainey Harper, 1891–1906. Office of the President, University of Chicago. Available at http://president.uchicago.edu/directory/william-rainey-harper.

William W. Watson, 92, Physicist who helped develop atom bomb. *The New York Times*, August 6, 1992.

Zinsmeister K. *Julius Rosenwald*. The Philanthropy Hall of Fame, The Philanthro-phy Roundtable. Available at http://www.philanthropyroundtable.org/almanac/great_men_and_women/hall_of_fame/julius_rosenwald.

Chapter 6

Hook S. Heresy, yes—But conspiracy no. *The New York Times Magazine*, July 9, 1950.

Kirk R. 1956. Conservative vs. liberal—A debate. *The New York Times,* March 4, 1956.

Schlesinger A Jr. 1956. Conservative vs. liberal—A debate. *The New York Times*, March 4, 1956.

Schlesinger A Jr. 1956. Liberalism in America: A note for Europeans. In *The politics of hope*. Riverside Press, Boston [1962]. Available at http://www.writing.upenn.edu/~afilreis/50s/schleslib.html.

Sunset Hill Farm County Park. Available at http://www.duneland.com/aer/parks/sunsethill.html and http://www.indianadunes.com/parks-and-recreation/parks-and-facilities.

The press: Enter Perspectives USA. *Time*, April 14, 1952.

Watson J. February 8, 1957 letter from Jean Watson to James D. Watson, Jr. James D. Watson Collection, CSHL Archives.

Chapter 7

Alfred Tissières. Available at www.itqb.unl.pt/cssi/Alfred.pdf.

Carradale House. Available at http://www.carradale.org.uk/house.html.

Erdem M, Rose KW. 2000/2. American Philanthropy in Republican Turkey: The Rockefeller and Ford Foundations. *The Turkish Yearbook* XXXI: 131–157.

Heise K. 1985. Harvey S. Olson, 77, author of best-selling travel guides. *The Chicago Tribune*, December 21, 1985.

Helen Gahagan Douglas. Women in Congress. Office of the Clerk U.S. Capitol, Room H154, Washington, DC 20515-6601. Available at http://womenincongress.house.gov/member-profiles/profile.html?intID=61.

Jones VC. 1985. *United States Army in World War 2, Special studies. Manhattan: The army, and the atomic bomb*, pp. 245–247. Government Printing Office, Washington, DC. Available at http://www.history.army.mil/html/books/011/11-10/index.html.

Lanouette W (with Szilard B). 1992. *Genius in the shadows: A biography of Leo Szilard, the man behind the bomb.* Macmillan, New York.

Mitchison A. 2003. Jim's cool reception among the British geneticists. In *Inspiring science: Jim Watson and the age of DNA.* Cold Spring Harbor Laboratory Press, Cold Spring Harbor, NY.

Mitchison D. Dick Mitchison. CSHL Archives Repository. Available at http://libgallery.cshl.edu/items/show/51802.

Mitchison N. Letters from Naomi Mitchison to James D. Watson. CSHL Archives Repository. Available at http://libgallery.cshl.edu/items/show/45471. CSHL Archives Repository. Available at http://libgallery.cshl.edu/items/show/45452.

Mitchison NA. Handwritten letter from Avrion Mitchison to James D. Watson. Available at http://libgallery.cshl.edu/items/show/45417.

Mitchison NA. Letter from Avrion Mitchison to James D. Watson. CSHL Archives Repository. Available at http://libgallery.cshl.edu/items/show/45415.

Mitchison Wedding, Isle of Skye. CSHL Archives Repository. Available at http://libgallery.cshl.edu/items/show/51584.

Myers EW. Copy of letter from Elizabeth Watson Myers to her parents and grandmother. CSHL Archives Repository. Available at http://libgallery.cshl.edu/items/show/45855.

Myers EW. Copy of letter from Elizabeth Watson Myers to James D. Watson. CSHL Archives Repository. Available at http://libgallery.cshl.edu/items/show/45859.

Myers EW. Letter from Elizabeth Watson Myers to James D. Watson. CSHL Archives Repository. Available at http://libgallery.cshl.edu/items/show/45816.

Neumode Hosiery Co., Chicago, Ill. 1951. US Patent Office Registration, no. 550,108.

Pauling L. Letter from Linus Pauling to James Watson and Francis Crick, March 27, 1953. Linus Pauling and the Race for DNA, Oregon State University. Available at http://osulibrary.orst.edu/specialcollections/coll/pauling/dna/corr/sci9.001.34-lp-watsoncrick-19530327.html.

The Salvador E. Luria Papers. Biographical information. Profiles in Science, National Library of Medicine. Available at http://profiles.nlm.nih.gov/ps/retrieve/Narrative/QL/p-nid/164.

Watson DC. Correspondence. CSHL Archives Repository. Available at http://libgallery.cshl.edu/items/show/50918. And CSHL Archives Repository. Available at http://libgallery.cshl.edu/items/show/50919.

Watson JD, Crick FHC. Prepublication typescript of "A structure for DNA," approximately March 21, 1953. Ava Helen and Linus Pauling Papers. Linus Pauling and the Race for DNA, Oregon State University. Available at http://osulibrary.orst .edu/specialcollections/coll/pauling/dna/notes/sci9.001.32-ts-19530321.html.

Watson JD. 1968. *The double helix: A personal account of the discovery of DNA.* Atheneum, New York.

Watson JD (Stent GS, ed). 1980. *The double helix: a personal account of the discovery of DNA.* A Norton Critical Edition. WW Norton & Company, New York.

Watson JD, Crick FHC. 1953. A structure for deoxyribose nucleic acid. *Nature* 171: 737–738.

Watson JD. Letter from James D. Watson to Max Delbrück, March 12, 1953. James D. Watson Collection, CSHL Archives, Cold Spring Harbor, NY.

Watson JD. Salvador Luria's biography by James D. Watson. CSHL Archives Repository. Available at http://libgallery.cshl.edu/items/show/43121.

Watson JD Sr. Correspondence. CSHL Archives Repository. Available at http:// llibgallery.cshl.edu/items/show/50929; http://llibgallery.cshl.edu/items/show/50930; and http://libgallery.cshl.edu/items/show/50932.

Watson JD Sr. 1959. Letters to Betty and Bob Myers. CSHL Archives, Cold Spring Harbor, NY.

Watson WW. Correspondence. CSHL Archives Repository. Available at http:// libgallery.cshl.edu/items/show/50937.

Watson WW. Letter from William Weldon Watson to James D. Watson. CSHL Archives Repository. Available at http://libgallery.cshl.edu/items/show/50938.

Chapters 8 and 9

AllPolitics. 1997. *Chicago '68. A chronology.* Available at http://chicago68.com/ c68chron.html.

American Masters: Walter Cronkite. Educational Broadcasting Corporation. http:// www.pbs.org/wnet/americanmasters/episodes/walter-cronkite/about-walter-cronkite/561/.

Bernstein A. 2010. Laughlin Phillips, 85: CIA officer and art museum chairman Laughlin Phillips, 85, dies. *Washington Post*, January 26, 2010.

Drabble M. The English degenerate [re: John Cowper Powys]. *The Guardian*, August 11, 2006.

Feynman RP. Handwritten letter from Richard Feynman to James D. Watson. CSHL Archives Repository. Available at http://libgallery.cshl.edu/items/show/36852.

Friedberg EC. 2005. *The writing life of James D. Watson.* Cold Spring Harbor Laboratory Press, Cold Spring Harbor, NY.

Goossens R. *Remembering the classic liners of yesteryear. M.S.* Oranje. Available at http://www.ssmaritime.com/oranje.htm.

Hopkins N. Correspondence. CSHL Archives Repository. Available at http://libgallery.cshl.edu/items/show/.

Hudson S. 1999. *Snow business: A study of the international ski industry*, pp. 44–46. Cassell, London.

Inglis JR, Sambrook J, Witkowski JA. 2003. *Inspiring science: Jim Watson and the age of DNA.* Cold Spring Harbor Laboratory Press, Cold Spring Harbor, NY.

Johnson LB. 1968. President Lyndon B. Johnson's address to the nation announcing the steps to limit the war in Vietnam and reporting his decision not to seek reelection, March 31, 1968. Public Papers of the Presidents of the United States: Lyndon B. Johnson, 1968–69, Vol. 1, entry 170, pp. 469–476. Government Printing Office, Washington, DC. Available at http://www.lbjlib.utexas.edu/johnson/archives.hom/speeches.hom/680331.asp.

Lewis RV. Invitation to Elizabeth Lewis and James D. Watson's wedding. CSHL Archives Repository. Available at http://libgallery.cshl.edu/items/show/47402.

Mallia N. 2013. Press at war: The Tet offensive as a turning point in Vietnam. *The US History Scene.* Available at http://www.ushistoryscene.com/uncategorized/pressatwar.

McKie R. 2012. DNA pioneer James Watson reveals helix story that was almost never told. *The Observer*, December 8, 2012.

Nancy H. Hopkins. The David H. Koch Institute for Integrative Cancer Research at MIT. http://ki.mit.edu/people/faculty/hopkins.

Pace E. 1998. John H. Richardson, 84, C.I.A. station chief in early '60s. *The New York Times*, June 14, 1998.

Penetrating heredity's secrets. *The New York Times*, October 19, 1952.

Pontecorvo G. Letter from Guido Pontecorvo to James D. Watson. CSHL Archives Repository. Available at http://libgallery.cshl.edu/items/show/48572.

Richardson J. 2005. The spy left out in the cold. *The New York Times*, August 7, 2005.

Sullivan W. 1968. A book that couldn't go to Harvard. *The New York Times*, February 15, 1968.

Watson JD Sr. Correspondence. CSHL Archives Repository. Available at http://libgallery.cshl.edu/items/show/50918; http://libgallery.cshl.edu/items/show/50919; http://libgallery.cshl.edu/items/show/50927; http://libgallery.cshl.edu/items/show/50931; and http://libgallery.cshl.edu/items/show/50932.

Watson JD Sr. Diary extracts, 1960–1967. CSHL Archives, Cold Spring Harbor, NY.

Watson WW. Correspondence. CSHL Archives Repository. Available at http://libgallery.cshl.edu/items/show/50937 and http://libgallery.cshl.edu/items/show/50938.

Webster B. 1980. Robert Ardrey dies; Writer on behavior. *The New York Times*, January 16, 1980.

Picture Credits

Chapter 1: **p. 1**, Collection of Dr. and Mrs. James D. Watson; **p. 3**, courtesy of Mary Coe Foran; **p. 4**, *Illinois Weekly State Journal*, 1836; **p. 5**, with permission from Sagamon County Historical Society; **p. 6**, U.S. Census Bureau; **pp. 7, 8**, courtesy of Benjamin A. Watson Gold Rush Letters, 1849–1851 Collection, California State Library; **p. 9**, *Daily Illinois State Journal*, Springfield, Illinois; **pp. 10, 12**, collection of Douglas Pepin; **p. 13**, *Harper's Weekly*, May 20, 1865 (reproduced with permission of HarpWeek); **p. 14**, reproduced from Ruger, Albert, "Springfield, Illinois 1867. Drawn from nature by A. Ruger," 1867. Library of Congress, Geography and Map Division; **p. 15**, reproduced from https://www.eacgallery.com/LotDetail .aspx?inventoryid=5397; **p. 16**, The Abraham Lincoln Papers at the Library of Congress; **p. 17**, reprinted from Pike County IL, combined 1872 and 1912 atlases (http://freepages.genealogy.rootsweb.ances try.com/!glendasubyak/platindex.html); **p. 18**, Illinois General Assembly, Laws of the State of Illinois; **p. 19**, courtesy of the James D. Watson Family Collection, James D. Watson Collection, Cold Spring Harbor Laboratory Archives; **p. 20, top**, courtesy of Quincy Public Library; **p. 20, bottom**, Historic Map Works LLC; **p. 21**, courtesy of the Pike County Historical Society; **p. 22, top left**, *Daily Illinois State Register*, Sunday, June 23, 1901; **p. 22, bottom left**, supplied from Ewell/Watson family archives by Dana Ewell; **p. 22, bottom right**, courtesy of Elizabeth L. Watson; **p. 23**, supplied from Ewell/Watson family archives by Dana Ewell; **p. 24**, U.S. Census Bureau; **p. 25**, published by Artistic Publishing Association, Boston, MA (http://www.steinerag.com/flw/Artifact%20Pages/PhRtS171whit ing.htm); **p. 26, top left**, *Daily Illinois State Journal*, February 27, 1882, Springfield, IL; **p. 26, right**, *Daily Illinois State Register*, March 25, 1896, Springfield, IL; **p. 27**, supplied from Ewell/Watson family archives by Dana Ewell; **pp. 28, 29, left**, supplied from Ewell/Watson family archives by Dana Ewell; **p. 29, right**, Chicago History Museum, DN-0003226m (original negative ichicdn n003226).

Chapter 2: **pp. 31–34, 38**, Courtesy of the James D. Watson Collection, Cold Spring Harbor Laboratory Archives; **pp. 39, left, 40**, courtesy of the Wilson Ornithological Society and its publications (*Wilson Journal of Ornithology* and *Wilson Bulletin*); **p. 39, right**, courtesy of Oberlin College Archives; **p. 41**, courtesy of Photo Archives Collection: RG 3.04, Presidents: William J. Hutchins, Berea College Archives, Special Collections & Archives, Hutchins Library, Berea College; **p. 43**, courtesy of Oberlin College Archives; **pp. 44, 45**, supplied from Ewell/Watson family archives by Dana Ewell; **pp. 46, 47, top**, courtesy of the James D. Watson Collection, Cold Spring Harbor Laboratory Archives; **pp. 47, bottom, 48–52**, courtesy of the James D. Watson Family Collection, James D. Watson Collection, Cold Spring Harbor Laboratory Archives.

Chapter 3: **p. 53**, Chicago History Museum reproduction from film, *The Kirtland's Warbler in Its Summer Home*: third known nest (1982.20); (n.p.); ca. 1923; Creator—James McGillivray; **p. 54**, Chicago History Museum; **p. 55**, Chicago History Museum, reproduction of Glass negative DN-0077059; Photographer-Chicago Daily News, Inc.; **pp. 57–60**, courtesy of the James D. Watson Collection, Cold Spring Harbor Laboratory Archives; **p. 61, top**, reprinted from the *Jack Pine Warbler*, Oct 1943, Vol 21, Issue 4, with permission from the Michigan Audubon Society; **p. 61, bottom**, photo by USFWS; Joel Trick; **pp. 63–64**, Chicago History Museum reproduction from film, *The Kirtland's Warbler in Its Summer Home*: third known nest (1982.20); (n.p.); ca. 1923; Creator—James McGillivray; **pp. 65–70**, reprinted from Leopold NF Jr. 1923. *The Auk* 40: 409–414, with permission from the American Ornithologists' Union; **p. 71, top**, reprinted from Leopold NF Jr. 1924. *The Auk* 41: 44–58, with permission from the American Ornithologists' Union; **p. 71, bottom right**, http://archive.org/details/iragionamenti02aretuoft; **p. 72**, Chicago History Museum, reproduction of Glass negative DN-0077041; Photographer-Chicago Daily News, Inc.; **p. 73**, Chicago History Museum, reproduction of Glass negative DN-0077257; Photographer-Chicago Daily News, Inc.; **p. 75**, Chicago History Museum, reproduction of Glass negative DN-0077611; Photographer-Chicago Daily News, Inc.; **p. 78**, from Clerk of the Circuit Court, Cook County, Illinois (http://www.cookcountyclerkofcourt.org/?section= RecArchivePage&RecArchivePage=leopold_and_loeb); **p. 79**, Clarence Darrow Digital Collection, University of Minnesota Law Library; **p. 80, top,** Chicago History Museum, reproduction of Glass negative DN-0077499; Photographer-Chicago Daily News, Inc.; **p. 80, bottom left,** Library of Congress, Prints & Photographs Division (GG Bain Collection 38189); **p. 81,** Chicago History Museum, reproduction of Glass negative DN-0078037; Photographer-Chicago Daily News, Inc.; **p. 82,** Chicago History Museum, reproduction of Glass negative DN-0078038; Photographer-Chicago Daily News, Inc.; **p. 83,** Clarence Darrow Digital Collection, University of Minnesota Law Library; **p. 84, left,** Chicago History Museum, reproduction of Glass negative DN-0078242; Photographer-Chicago Daily News, Inc.; **p. 84, right,** Chicago History Museum, reproduction of Glass negative DN-0078243; Photographer-Chicago Daily News, Inc.; **p. 85,** Book Cover, ©1958 Doubleday; from *Life Plus 99 Years* by Nathan F. Leopold. Used by permission of Doubleday, an imprint of the Knopf Doubleday Publishing Group, a division of Random House LLC. All rights reserved; **p. 86, top,** Agricultural Experiment Station, University of Puerto Rico-Mayagüez; **p. 86, bottom,** ©1959, 20th Century Fox Corp.

Chapter 4: **p. 89,** Courtesy of the James D. Watson Collection, Cold Spring Harbor Laboratory Archives; **p. 90, bottom left,** collection of Dr. and Mrs. James D. Watson; **p. 90, top,** courtesy of the James D. Watson Family Collection, James D. Watson Collection, Cold Spring Harbor Laboratory Archives; **p. 91,** courtesy of the James D. Watson Family Collection, James D. Watson Collection, Cold Spring Harbor Laboratory Archives; **p. 92,** courtesy of the James D. Watson Collection, Cold Spring Harbor Laboratory Archives; **p. 93,** courtesy of the James D. Watson Collection, Cold Spring Harbor Laboratory Archives; **p. 94,** courtesy of the James D. Watson Family Collection, James D. Watson Collection, Cold Spring Harbor Laboratory Archives; **p. 95, top left,** from website no longer available; **p. 95, top, bottom right,** courtesy of the James D. Watson Collection, Cold Spring Harbor Laboratory Archives; **pp. 96–97,** courtesy of the James D. Watson Collection, Cold Spring Harbor Laboratory Archives; **p. 98,** U.S. Census Bureau; **p. 99,** courtesy of the James D. Watson Collection, Cold Spring Harbor Laboratory Archives; **p. 101,** from Lori Ferber Collectibles (loriferber.com); **p. 102, bottom left,** courtesy of Catherine (Olvaney) Stelleman; **p. 102, top right,** courtesy of the James D. Watson Collection, Cold Spring Harbor Laboratory Archives; **pp. 104–109, 111–112,** courtesy of the James D. Watson Collection, Cold Spring Harbor Laboratory Archives; **p. 113,** Schlesinger Library, Radcliffe Institute, Harvard University. Photo by Jessie Tarbox Beals; **p. 114,** courtesy of the James D.

Watson Collection, Cold Spring Harbor Laboratory Archives; **p. 115,** supplied from Ewell/Watson family archives by Dana Ewell; **p. 116,** archives of the Woodstock, Illinois Public Library; **p. 117,** Library of Congress, Prints & Photographs Division, Carl Van Vechten Collection [reproduction number, e.g., LC-USZ62-54231.

Chapter 5: **p. 119,** Courtesy of the James D. Watson Family Collection, James D. Watson Collection, Cold Spring Harbor Laboratory Archives; **p. 120,** William Rainey Harper College Archives; **pp. 121, 123–124,** with permission from Special Collections Research Center, University of Chicago Library; **p. 125,** with permission from Special Collections Research Center, University of Chicago Library; MagnumPhotos, New York, photograph by Philippe Halsman; **pp. 127, 128,** courtesy of the James D. Watson Collection, Cold Spring Harbor Laboratory Archives; **p. 129, top,** Time & Life Pictures/Getty Images; **p. 129, bottom,** Wikipedia; **p. 130,** with permission from Special Collections Research Center, University of Chicago Library; **p. 131, top right,** Wikipedia; **p. 131, bottom left,** U.S. Department of Energy (DOE), courtesy AIP Emilio Segre Visual Archives; **pp. 133–137,** courtesy of the James D. Watson Family Collection, James D. Watson Collection, Cold Spring Harbor Laboratory Archives; **p. 138,** with permission from Special Collections Research Center, University of Chicago Library; **p. 139,** courtesy of the James D. Watson Collection, Cold Spring Harbor Laboratory Archives; **pp. 140–150,** courtesy of the James D. Watson Collection, Cold Spring Harbor Laboratory Archives.

Chapter 6: **p. 151,** Courtesy of James D. Watson Collection, Cold Spring Harbor Laboratory Archives; **p. 152,** Steven R. Shook Collection.

Chapter 7: **p. 165,** Courtesy of the James D. Watson Collection, Cold Spring Harbor Laboratory Archives; **p. 166,** collection of Dr. and Mrs. James D. Watson; **p. 167,** courtesy of the James D. Watson Collection, Cold Spring Harbor Laboratory Archives; **p. 168,** Harvard University Archives, Papers of James Dewey Watson, 1945–1968, HUG 4874.1404; **pp. 169–173,** courtesy of the James D. Watson Collection, Cold Spring Harbor Laboratory Archives; **p. 175, top,** courtesy of Cold Spring Harbor Symposia on Quantitative Biology Collection, Cold Spring Harbor Laboratory Archives; **p. 175, bottom right,** Cold Spring Harbor Laboratory Archives; **p. 176,** courtesy of the James D. Watson Collection, Cold Spring Harbor Laboratory Archives; **p. 177, top,** from the collection of Lisanne F. Anderson, with permission; **p. 177, bottom,** courtesy of the James D. Watson Collection, Cold Spring Harbor Laboratory Archives.

Index

Page references in *italics* refer to information found in the annotations and figure legends.